They Work with Wildlife

Also by Edward R. Ricciuti

Plants in Danger
Shelf Pets: How to Take Care of Small Wild Animals
Sounds of Animals at Night
To the Brink of Extinction

THEY WORK
WITH
WILDLIFE

JOBS FOR PEOPLE
WHO WANT TO WORK
WITH ANIMALS

Edward R. Ricciuti

1 8 17
HARPER & ROW, PUBLISHERS
Cambridge, Philadelphia, San Francisco, London, Mexico City, São Paulo, Sydney
NEW YORK

They Work with Wildlife: Jobs for People Who Want to Work with Animals
Copyright © 1983 by Edward Raphael Ricciuti

Library of Congress Cataloging in Publication Data
Ricciuti, Edward R.
    They work with wildlife.

    Summary: Introduces various jobs which involve
working with wildlife, including game warden, nature
writer, and field biologist.
    1. Animal specialists—Vocational guidance—Juvenile
literature. 2. Zoologists—Vocational guidance—
Juvenile literature. [1. Animal specialists—Vocational
guidance. 2. Zoology—Vocational guidance. 3. Vocational
guidance] I. Title.
QL50.5.R53 1983   591'.023   80-7918
ISBN 0-06-025003-8
ISBN 0-06-025004-6 (lib. bdg.)

Designed by Trish Parcell
10 9 8 7 6 5 4 3 2 1
First Edition

For William Conway, worker with wildlife

Contents

vii

They Work with Wildlife

Introduction: Do You Want to Work with Wildlife?

Because I write about nature and have been a curator for the New York Zoological Society, people often ask my advice about how they can find jobs enabling them to work with wild animals.

"I'm interested in conservation," they often say. "I want to help wildlife. Animals appeal to me, and I can't think of anything that would be more fun to do for a living than work with them. It would never be boring. It would be exciting. How do I do it?"

If these people are adults, set in their ways and careers, perhaps with family responsibilities, they may receive a discouraging answer. Unfortunately, for many of them it is too late in life to begin a new career working with animals, at least unless they make drastic changes in their ways of life, which many cannot. They lack the experience and training needed for the jobs that attract them so much.

"If only I had known earlier" is the comment often made after I explain some of the qualifications needed to be a marine biologist, for instance, or a zoo keeper. Young people, on the other hand, have the chance to prepare for careers with wild animals. There are many various jobs that can bring you in touch with wildlife. This book describes some of them—and shows what some people did to enable them to work with wildlife.

1

Wildlife Management

Throughout the north country, the big Canada geese are growing restless. A deep uneasiness seems to have seized both the adult geese and their young of the year, recently grown to full size. Honking, small flocks take to the air, circle, and then land again. Thousands upon thousands of the black-headed, white-cheeked birds, which have been in the region around Hudson Bay since spring, are preparing—although by instinct, not conscious planning—for their yearly migration south.

Each day the geese become more restless, until the first flocks depart. As they have countless times before, the birds leave the ground and circle, but this time they do not return. Instead, wings steadily beating, they head south. Day after day, others follow, and finally the last are airborne. When the echoes of their haunting calls fade from the northern skies, the northland will be without Canada geese until the next spring, when the flocks return.

Huge birds, the geese fly southward in great V-shaped formations. Even though the days of early autumn are still warm, the wedges of geese are a sure sign to observers on the ground that the cold weather is not really very far away.

By the time winter arrives, however, the geese are at home in their southern destinations, where the weather is not as severe and food is abundant. Geese from eastern Canada winter in several places along the middle Atlantic and southeastern coasts of the United States, but especially on the Eastern Shore of Chesapeake Bay, in Delaware and Maryland. More of the big "honkers" winter here, in fact, than any other place in North America. The reason is that the Eastern Shore is a major farming area, with vast fields of soybeans, corn, and other crops on which geese feed. After the fall harvest, enough remains on the ground to keep the birds fat, so the region can support immense numbers of the birds.

I have watched a tiger walk the forest of southeastern Asia, a condor sail over the peaks of South America's Andes, hippos snorting in African rivers, and reef fish under the Red Sea, but nothing I have seen in nature is any more astonishing than the gathering of geese on the Eastern Shore.

Huddling in the dark just before daybreak, I have

waited for the vast flocks. As the sun begins to rise, the sound of honking drifts in ever so faintly from beyond the horizon. Slowly the volume increases until it seems to fill the world. Thousands of geese call together, and soon the air seems to vibrate with sound. When the flocks appear, it can be almost as if night is returning. They darken the sky, coming on in great waves, flock after flock seeming without end.

A Place for Geese

Because the goose population is large and healthy, geese can be hunted for sport. Waterfowl hunters from all over the country come to the Eastern Shore after geese, both on private land and on lands administered by state and federal governments. The geese did not always winter on the Eastern Shore in such large numbers. As open farm fields replaced woods over the area, however, the habitat—the environment that suits an animal—improved for geese. They began to gather along the outer shore of Chesapeake Bay.

Today, farmers and other landowners encourage the geese to use the region by planting food and digging ponds for them, so they may lease hunting rights to sportsmen. The state fish and game departments, and the United States Fish and Wildlife Service, operate refuges and preserves on land where the habitat has been developed for geese. All of these activities are meant to insure that the geese will prosper.

Wildlife Management

Planting crops and creating open water for geese are part of a profession called *wildlife management*. In broad terms, wildlife management is the effort to promote natural resources such as soil, water, and vegetation in the interests of wildlife and, in the long run, for a better environment.

You might ask why wild animals are not simply left alone to survive, as they have in the past without human help. The answer is that no part of nature today, even in remote places, has not been influenced by human activities.

In the developing countries of the tropics, for instance, human numbers are exploding and people are flooding into forests and onto plains that once were remote wildernesses. Wildlife there is losing its habitat.

Air pollution from industry in the United States, to use another example, is causing conditions in the atmosphere that create acid rain. The polluted rain falls from clouds carried by winds to the north, and poisons lakes deep in the Canadian wilderness. Fish and other water animals die.

Wildlife management tries to offset the damage done by people, and to influence conditions in favor of wild animals. As a profession in the United States, wildlife management is less than a century old. In the early 1900s, conservationists feared that most large animals in the United States would soon disappear. But they did not figure on the increasingly scientific methods by which wildlife managers were to operate.

Their achievements have been astonishing. To name just a few: In 1895 there were 350,000 white-tailed deer below the Canadian border, and they were gone from half the states. Today due to wildlife management there are 12 million whitetails in the United States. The wild turkey was gone from the majority of the states and common only in a few parts of the south. Now almost all the states have wild turkey populations. In 1915 the wood duck was almost extinct. Today it is the most common waterfowl breeding in the eastern United States.

Saving and restoring animals like these involves many activities. The animals must be protected by law to insure that the few remaining are not killed off. Habitat must be set aside and improved so the creatures flourish. Once the numbers of the animals begin to build, new habitat must be created, usually in areas the creatures once lived in but have disappeared from.

In the case of wood ducks, for instance, wildlife managers set up nest boxes to replace the hollow trees in which the ducks like to nest. The trees had been destroyed as woodlands were cleared. The ducks favor swamps as habitat. Many woody swamps in which they once raised their young and fed had been drained. Wildlife managers set up small dams along streams through wooded areas and created new swamps for the ducks. Today wood ducks, like whitetails and wild turkeys, are numerous enough to be hunted—under rules established by wildlife managers—without their species being endangered.

Education and Employment

Wildlife managers need a solid understanding of biology, ecology, and related sciences. College, therefore, is essential for anyone who is looking for a job in the field. A major in the life sciences—some schools even have undergraduate courses in wildlife management—is perhaps the best.

Most wildlife managers, moreover, have master's degrees in their field or a similar area. Some have doctorates. Those with advanced degrees find that doors open more easily when they are seeking jobs, and that they also have a wider choice of positions.

Wildlife managers can wind up in a variety of posts. Some manage refuges and sanctuaries. Others work as consultants, hiring out to various employers for particular projects. Wildlife managers help government agencies establish regulations on hunting—how many animals of a particular species in a certain area may be taken without harming the species, for example.

Several federal agencies employ wildlife managers. These include the Fish and Wildlife Service, the Forest Service, the National Park Service, the Bureau of Land Management, the Environmental Protection Agency, and even the armed forces, which use wildlife managers to care for animals on large military bases. The states' departments of conservation, fish and game, and environmental protection all have wildlife managers.

Aside from the government, almost any organization or business with areas of wildlife and habitat under its control may need a wildlife manager to look after things. Among these are nature sanctuaries, timber companies, ranches, and game farms.

Remington Farms

One of the best-known centers of wildlife management in the United States is Remington Farms. It is located in the midst of goose country on the Eastern Shore of Chesapeake Bay, at Chestertown, Maryland, and is owned by the Remington Arms Company of Bridgeport, Connecticut.

The company makes firearms and ammunition for sportsmen, and therefore encourages the management of game animals. At the same time, however, Remington is interested in promoting wildlife conservation in general, and has management programs for animals not sought by hunters as well as for game.

Remington Farms covers more than 3,000 acres, a third of which can be tilled for crops. The rest is woodlands, marsh, and water. There are 26 freshwater ponds, varying in size from one to 50 acres. Properly run, farms can be excellent wildlife habitats, particularly for creatures like white-tailed deer, waterfowl, pheasants, and many songbirds. The aim of Remington Farms is to demonstrate ways in which farming and wildlife can benefit one another: how to manage wild animals on agricultural lands so that the farmer gets the most out of his crops and also has wildlife to enjoy.

A real working farm, Remington has equipment,

barns, sheds, and houses for employees. The farm produces corn, soybeans, timber, and other crops just like those grown on the ordinary type of farm. Agricultural production at Remington is carried on according to the latest techniques of soil and water conservation. In general, it is a model farm for both people and wildlife.

Among the wild animals living on Remington Farms are deer, rabbits, foxes, beavers, raccoons, herons, egrets, a variety of ducks, and Canada geese, which are the main attraction. All told, about 20,000 Canada geese live there from fall to early spring. They are birds that spend the warm months on breeding grounds in Canada, around Hudson and James bays and through Quebec to the Atlantic coast.

The Outdoor Life

Two wildlife managers run the Remington Farms operation. E. Hugh Galbreath is head manager. Edward C. Soutiere is his assistant. While people in other wildlife management posts may have somewhat different responsibilities, following Galbreath's and Soutiere's careers can give you a sound idea of what it is like to manage wildlife.

Galbreath and Soutiere work as a team overseeing the farm and projects carried out there. Both have had many years as wildlife managers, but neither had boyhood dreams about entering the field. Each, however, eventually decided he would like to work outdoors, and found that wildlife management was the way to do it.

Soutiere was raised in Vermont, where he learned to hunt and fish. Galbreath is from a small town in southeastern Georgia. He hunted squirrels and fished with his father but, he says, "I certainly could not say I was an avid hunter or fisherman." He merely liked being outside.

Galbreath's outdoor interests led him to thinking about wildlife and forestry. He decided to attend the University of Georgia, which offered wildlife courses in its School of Forest Resources.

Soutiere developed a keen interest in science during high school, because, he says, "I had excellent science teachers." The problem was, Soutiere remembers, "I didn't know what kind of scientist I wanted to be."

After a try at college, Soutiere quit and joined the United States Air Force. He was given an office job, which decided him on his future. "I discovered I did not want a job that kept me indoors," he says. When he was discharged from the military, he enrolled at the University of Vermont to study wildlife.

Soutiere's undergraduate degree was in forestry. To this he added a master's degree and doctorate in range management and wildlife management, plus a stint as a wildlife biologist for the state of Iowa. Then he joined Remington Farms.

Galbreath had hands-on experience in wildlife management while still in college. He worked the summers before his junior and senior years as a student trainee at the Santee National Wildlife Refuge, operated by the Fish and Wildlife Service at Summerton, South Carolina.

"That was good experience and helped me decide on

wildlife management as a career," Galbreath says, "but I suspect that at least half the student trainees changed their majors after they were exposed to what wildlife management was really all about."

After graduating from college, Galbreath worked for a summer at another national wildlife refuge, Loxahatchee, in the Everglades of Florida. Then he returned to Georgia for a master's degree. His next position was at the Savannah National Wildlife Refuge, an area of vast marshes filled with water birds and waterfowl. There, working under the refuge manager, he served an apprenticeship.

"It was a very good experience, because the manager gave me some responsibility and allowed me to make some mistakes," Galbreath says.

Galbreath's next job was as an assistant county agricultural agent in La Grange, Georgia. He helped farmers and others manage and develop their lands and provided information on managing natural resources to schools and citizens groups. While Galbreath was attending the wedding of a friend, he heard from another wildlife manager that an assistant manager position was open at Remington Farms. Galbreath applied in 1967 and was given the job. Ten years later the manager retired, and Galbreath succeeded him.

Working at Remington

Galbreath's experience is invaluable to him as the top man at Remington Farms. While at national wildlife refuges he learned management operations from the ground up. As a county agent he had to communicate with people about managing resources. Today, besides taking care of his management duties, he talks to school children and civic clubs about wildlife management, and oversees the publication of pamphlets and production of films about the farm and projects there.

Wildlife managers, Galbreath stresses, do not spend all their time with animals. Galbreath must administer the farm, planning budgets for staff and equipment, overseeing the building of roads, and creating an annual plan for activities at the farm

Long-range planning begins in midwinter. With his staff Galbreath decides on actions such as what crops should be planted and where, how to improve wildlife habitat in various sections of the farm, whether timber should be harvested, and whether wood ducks on farm ponds need new nest boxes. What is the expected corn yield? These and many other questions must be examined.

Soutiere helps in planning, but is less concerned with administration than his supervisor. He is in the field more but, he notes, "I still have to keep records and write reports. It's not all outdoor work."

One of Soutiere's responsibilities is overseeing maintenance of the farm's buildings, making sure they are kept in good repair. He may start his workday by making up a work schedule for the men who perform building maintenance, keeping track of details such as whether the rest rooms are operating or a new door must be ordered.

Next, he might review a wildlife plan for land owned by a local farmer or the state. As a public service, Soutiere and Galbreath help landowners and the government develop wildlife habitat on their properties. They visit the land in question, look over it and its wildlife, and then suggest how the habitat may be improved.

Before starting a wildlife management plan, Soutiere obtains maps showing the land, its terrain, and soil and water resources. He may also need aerial photographs of the property so he can view it both as a whole and in detail. Among the questions he must answer is what type of animals does the land—its vegetation, water, and other attributes—best support? Deer? Quail? Squirrels? Waterfowl? Once such questions are answered, Soutiere can draw up a management plan for the landowner to follow.

Along with Galbreath, Soutiere escorts visitors around the farm and explains what goes on there. Remington hosts about 50 groups a year, including students, businessmen, and biologists. "A wildlife manager in my position must be able to speak before other people," Soutiere explains.

During the course of his day, Soutiere may mark timber that is to be cut, or designate areas where shrubs should be planted to provide food and cover for songbirds. He also spends part of his time teaching wildlife management to graduate students. Each year Remington provides a scholarship to a graduate student, enabling him or her to work at the farms under the staff there for a year. The students assist Soutiere and Galbreath in various projects.

One of these projects is a study of the endangered Delmarva fox squirrel. This squirrel once ranged from Pennsylvania through Delaware and Maryland to Virginia. Its stronghold was the Eastern Shore, sometimes called the Delmarva Peninsula because *Del*aware, *Mary*land, and *Virginia* meet there. The fox squirrel, larger than the common gray squirrel, now numbers perhaps a few hundred animals. It declined because the forests in which it livod were cut. Soutiere is studying the creature in hopes of developing ways to manage habitat that will favor it even though the areas in which it lives arc no longer covered by deep forest.

Delmarva fox squirrels once lived in the neighborhood of Remington Farms, but they vanished long ago. In 1980 eight squirrels were captured on the nearby Blackwater National Wildlife Refuge, onc of the creature's last havens, and released on Remington Farms. The squirrels have reared many young.

Soutiere has observed the squirrels year round, especially during January and August, their breeding seasons, in hopes of learning as much as possible about their habitat needs. He catches the creatures and fits them out with radio transmitters, then releases them. By listening on a receiver for signals broadcast by the transmitters, Soutiere can follow the squirrels as they range the woodlands. To pinpoint the exact location of an animal, Soutiere works with a graduate student or other assistant, who also has a receiver but is located in a different position from the manager. By taking bearings on the squirrel from two positions, they can locate its whereabouts.

Although the fox squirrel looks and behaves something like the gray squirrel, there are differences in the ways the two species feed, Soutiere explains. The gray squirrel feeds mostly while in the branches. The Delmarva fox squirrel eats mainly on the ground, where it looks for fallen nuts and acorns and buds. This means that for a woodland to be a good habitat for the fox squirrel, the ground must be fairly open, not overgrown with brush that will get in its way or enable enemies to sneak up on it.

The welfare of a single species of squirrel may not seem to be an earthshaking affair. But the wildlife managers at Remington Farms look on reconciling the use of the land by people while keeping it suitable for wildlife as an important mission. Perhaps a booklet published by Remington explains it best:

Remember wildlife's needs are ours; food, water, shelter, and living space. And their requirements need not be sacrificed for ours, nor ours for theirs. It is not a case of man versus wildlife. We can live together. And our lives will be richer for it. If man is not compatible with wildlife, how can he be compatible with his own kind?

2

Field Biology

Great, gray hulks, elephants were feeding all around the four-wheel-drive vehicle in which I sat. Some of the huge beasts were almost within reach of my hand through the open window. Perhaps a hundred in all, the elephants tore up huge swatches of grass and low brush in their trunks and stuffed it into their mouths. A large bull eyed the vehicle for a moment and warily spread his huge ears, but then shook his head and returned to a tree from which he was trying to strip bark.

I was on the fringes of a deep, green swamp in the Amboseli National Park of Kenya, East Africa. Lying almost at the foot of snowcapped Mount Kilimanjaro, at 19,340 feet the highest peak in Africa, Amboseli is the home of a wealth of animals, including lions, leopards, black rhinoceroses, buffalo, and about 600 elephants.

The elephants of Amboseli have been studied closely for more than ten years by an American researcher who first went to East Africa in 1967 and fell in love with it. Cynthia Moss, in her early 40s, has lived much of the time virtually amidst the elephants. Her tent camp is at the edge of a swamp where the rich greenery regularly draws the giant creatures. Every now and then, one or more of the elephants visit the camp, often without incident but sometimes causing problems, such as wrecking her cook tent.

In fact, an elephant was dusting itself within a few yards of where we talked when I visited Moss at her camp. I had met her some years before, briefly, but now wanted to learn more about her research and how she did it.

What Is Field Biology?

Moss's research is what generally is called *field biology.* As I use it here, field biology really is a catchall category. It can be tied to a large number of scientific specialties, including wildlife biology, animal behavior, plant ecology, botany, and marine biology. Basically, however, it involves the study of living things in their native habitats to further understanding of how they exist.

The subject matter of field biologists is almost as varied as life on earth. One field biologist, for example, may track the migrations of whales across the ocean. Another may be interested in the feeding habits of bats. Still another may want to know more about the conditions that prompt certain monkeys to raid the crops of farmers, or why a rare insect is disappearing from its habitat.

Besides building our knowledge of nature, field biology can have a very practical, workaday goal. In today's world there is very little true wilderness. One way or another, human activities affect almost all of nature. Even as simple an activity as gathering firewood can have drastic consequences, especially in regions where human populations are rapidly exceeding the resources people need for living. In far corners of Asia's Himalayan Mountains, for instance, almost all the trees and brush have been stripped from the hillsides. Without vegetation to hold soil in place, it is carried off by rain and wind. What was once a rich habitat for plants and animals can become a wasteland.

Wildlife conservation in a changing world, in fact, is the main reason for most research in field biology today. Field biology can be the first step in managing the environment for wildlife. While most field biologists do not take part in actually working on the land like wildlife managers, the information they provide is extremely important in developing management plans. Some managers, moreover, may also carry on field research, as Edward Soutiere has on the fox squirrel at Remington Farms.

Cynthia Moss's observations, for example, are providing a wealth of interesting information on the way elephants live, and also what they need for survival—how much land, for instance, groups of various sizes require for enough food and space. Once such things are better understood, steps can be taken to offset or even avert human activities that could destroy elephant habitat.

Several types of organizations employ field biologists. These include government agencies like the Fish and Wildlife Service and state fish and game departments. Some large conservation organizations, such as the National Audubon Society, have field biologists on their staffs. One of the major conservation groups supporting field biology is the New York Zoological Society, famed for running the Bronx Zoo.

The society has a division called the Animal Research and Conservation Center. It has staff biologists who undertake projects as varied as studying the ecology of monkeys in Central Africa and following the behavior of whales off South America. Moreover, it provides money for research by other biologists in dozens of countries. Moss is one of them.

Over the years I have accompanied biologists afield in many parts of the world, observing them as they worked. On the plains of western South Dakota I watched Conrad N. Hillman, of the Fish and Wildlife Service, scan prairie dog towns with binoculars in search of the rare black-footed ferret, a member of the weasel family that probably is the mammal closest to extinction of any in North America. Ferrets live in empty prairie dog burrows and feed on prairie dogs, rodents somewhat resembling gophers.

Deep in the rain forest of the Luquillo Mountains of Puerto Rico, I walked through the jungle dimness with Cameron Kepler, a colleague of Hillman's at Fish and

During his studies of the Puerto Rican parrot, Kepler had to hike over rugged mountains and into jungle-covered valleys as difficult to travel as any landscape on earth. In such places, a slip or misstep can mean a broken bone or worse. The forest inhabited by the parrots, moreover, is hot and muggy, often swept by drenching downpours. It may be a beautiful place but often is an unpleasant one.

In search of the black-footed ferret, Hillman spent countless days and nights on the lonely, dark prairie in a vehicle. His watch continued through bitter-cold winters, when the wind that sweeps over the prairie can chill to the bone, and in the searing sun of the summer. Hillman had an advantage, however, because he grew up a North Dakota farm boy, lean and tough, a hunter of wild turkeys and coyotes, used to the tough conditions of the Dakota plains.

A large part of the field biologist's time, moreover, is spent waiting for things that never happen or poring over small details again and again. Imagine spending night after night in the cold, waiting to see a ferret that never appears. Or watching the same animal—any animal—do the exact same things for days, then weeks, and perhaps months on end. Every detail, furthermore, must be written down and later compared and totaled with sometimes hundreds of others.

Most field biologists have solid educations in zoology, ecology, animal behavior, or similar sciences. Field biology is a growing area of research but not a large one, says Archie Carr III, assistant director of the Animal Re-

Wildlife, as he showed me where he observed the endan
gered Puerto Rican parrot. Kepler trekked through th
forest, ever watchful for birds, counting them so the tot
number—it was once down close to only a dozen—coul
be known. He organized teams of volunteer bird watcl
ers to help in his parrot census. At the same time, Kepl
observed the birds in their nests, within hollow tree
and noted how events such as storms and heavy ra
affected them. Patiently, he watched rats, mongoose
hawks, and other creatures suspected of attacking pa
rots and their young, and found how much each of the
threatened the rare birds.

Eventually Kepler headed an effort to capture some
the parrots and keep them in captivity, where, hop
fully, they could be bred. This is just what has happene
at an aviary station set up in the same forest where t
parrots live in the wild. Other biologists have taken ov
the project, and the parrot seems on the way back fr
extinction.

Field biologists often work in exciting places a
sometimes experience adventures and sights most p
ple only dream about. Some field biologists become w
known all over the world through television progra
and books about their work. Make no mistake about
however, a field biologist pays for the excitement a
glamor with a body tired from hard, tough work, a
sometimes disease contracted while eating poor food
coming into contact with unsanitary conditions. O
sionally there is a brush with death—from an angry
phant, for example, or grizzly bear.

search and Conservation Center. Carr, who has a doctoral degree in wildlife ecology, knows well what it takes to be a top field biologist. He coordinates and monitors the activities of biologists from one end of the globe to another. And his father, Archie Carr II, is a biologist at the University of Florida, who has gained worldwide recognition for his field studies on behalf of wildlife conservation, especially helping to conserve the rare green sea turtle.

"The more education a field biologist has," Carr says, "the more doors open up when it comes time to find a job." Carr stresses that graduate school, for either a master's degree or a doctorate, is a great help. Not only does graduate study increase a person's educational qualifications, he notes, but it also offers a student the chance to work in the wild with established biologists. Field research is usually part of a graduate student's course of study. Experience actually working afield makes a person more valuable, says Carr, and also helps him or her discover what types of jobs are available in the first place.

Field biologists with master's degrees often find employmentment with government agencies. Those with doctorates have better chances of landing the most sought-after posts, such as field research in far places. Conrad Hillman, when I accompanied him onto the prairies, had obtained a master's degree. Cameron Kepler had his doctorate.

Cynthia Moss, on the other hand, has neither. She is most unusual in this respect, although even she admits that "people shouldn't get their hopes up" about becoming field biologists without scientific education. In Moss's case her dedication, perseverance, and a little luck made her dream of studying African animals in the wild come true.

From Newsweek *to Africa*

Moss was able to combine several of her qualifications and abilities into a package that eventually—but only after a difficult struggle—enabled her to find support for field research on African elephants. She was born in Ossining, New York, in Westchester County, outside New York City. From childhood on she had a deep interest in nature and later in wildlife conservation.

At college, however, she studied philosophy, in which she earned her undergraduate degree. A friendly, blond woman with a ready smile, Moss found a job as an editorial researcher for *Newsweek* magazine in New York City. While there she worked for *Newsweek*'s religion and theater departments, writing some reviews of plays.

All the while, she retained her liking for nature. She belonged to the Sierra Club, an organization dedicated to environmental conservation, and in 1967 took a vacation in East Africa to see some of that region's wildlife sights. It was then her big break occurred. Some friends, knowing of her interests, suggested she get in touch with Iain Douglas-Hamilton, a young British field biologist studying elephants in East Africa. Douglas-Hamilton has since become known around the world for his work, has been featured on television, and is a respected author.

He has pioneered new ways of observing elephants, living literally among them, observing them more closely perhaps than any scientist before. In the course of his work the biologist—with a reputation for fearlessness—has contracted serious disease, and been charged by elephants and trampled by a rhinoceros. All the while, he has continued to gain great insights into the way elephants behave toward one another, and how elephants live together—in effect, what goes into the making of elephant "society."

Moss spent a short time with Douglas-Hamilton in the field, helping him with some of his basic research tasks. She did such a good job he offered her a position as his research assistant. Although she returned to the United States for a few months, she accepted the offer and returned to Africa.

After working with Douglas-Hamilton as an apprentice, Moss was ready to start out on her own. Easier said than done. The road was extremely difficult, and many people would have been discouraged by the obstacles Moss faced. Once you get to know her, however, you realize that her interest in elephants and her dedication to learning about them are boundless. She needed all the help she could get. Field biologists do not simply go off into the wild and begin to look at animals. First they must think up a particular subject for their research. It must be worthwhile in terms of what it can do to increase scientific knowledge and conservation. It must not repeat what another biologist already has done.

Next there is the matter of money. Even a biologist who is already on the staff of an institution, and receives a salary, needs additional money to pay for research costs. Scientific research in the field, especially in out-of-the-way places, costs a great deal. Someone has to pay for air and ground transportation, equipment, food, and long lists of other items.

The money usually comes from government agencies or institutions such as foundations, which have funds for specific research projects. Competition among field biologists for these funds is heavy. Only those biologists with solid reputations and projects that are carefully thought out, and who have meaningful goals, are likely to obtain funds, although sometimes biologists with influential friends can obtain financial support even if their projects are not as worthwhile as others.

After finishing up her apprenticeship, Moss still lacked support for research of her own—or for that matter for herself. She took various jobs working in one way or another with animals. For a while she was employed at an orphanage for young wild animals, cleaning cages and caring for infant cheetahs, leopards, and similar creatures found abandoned in the wild.

Moss's editorial background with *Newsweek* also helped her earn a living while remaining in Africa. The producer of a film on elephants hired her to carry on background research. She earned additional money writing free-lance nature stories, and signed a contract with a publisher to write a book on research by field biologists in East Africa. An American-based conservation organization that supports conservation education in Africa, the African Wildlife Leadership Foundation, engaged Moss to edit its newsletter.

Meanwhile, as the opportunity arose, she assisted established field biologists with their research, although sometimes her only pay was experience. In 1972, on her own and without funds for her research, she began to study the elephants at Amboseli whenever she could find enough time. Within a few years the scientific world began to take note of what Moss was learning at Amboseli, and gradually funds to support her work began to come in. Her chief sponsors have been the African Wildlife Leadership Foundation and the New York Zoological Society.

Amboseli Elephants

Moss chose Amboseli as a site for studying elephants because although the area is a national park visited by mobs of tourists, the elephants there have hardly been disturbed by human activity. In many other parts of Africa elephants live under conditions that are unnatural. Farms have hemmed in many of the national parks where elephants live. The big animals no longer can wander far out of the parks as they once did. In other areas elephants have been totally eliminated to make way for farms and civilization. And through much of Africa poachers who kill elephants illegally for their ivory are a real threat. The poachers take as many elephants as they can, whereas legal hunters kill a limited number of animals, so the species is not endangered.

At Amboseli, however, things are different. The region, with its green swamps, dry plains, and scattered woodlands, is the home of the Masai tribe. They are tough warrior herdsmen who keep cattle but have no liking for farms. Elephants can range out of the boundaries of Amboseli as they did two and three hundred years ago, long before the park was established. The Masai, moreover, are not ivory poachers, nor will they tolerate people from other areas who attempt to kill elephants at Amboseli. For these reasons the Amboseli elephants are in as natural a state as can be found in East Africa, ideal for learning about genuine elephant behavior.

Camping with Elephants

When I visited Moss in Amboseli she was sharing her three-tent camp with another American field biologist, Phyllis Lee of Stanford University. Lee was studying the behavior of the vervet monkeys that scamper about the ground and trees in the neighborhood of the camp. An elderly African helped with the cooking. Next to the dining tent is their kitchen, built of large sticks, where the scientists have a gas stove for cooking and a portable refrigerator.

Set amidst a grove of palm trees, the camp is visited regularly by the monkeys, an occasional elephant, and once in a while some of the other beasts that live in Amboseli. Elephants have torn down the kitchen in search of something different to eat, and lions sometimes chew on the tent ropes after dark.

It sometimes pays to be watchful, says Moss, when going out of the tent at night. Once she left her tent and stumbled on a lion staring at her from only a few yards away.

As we walked around the camp and talked, I could hear elephants not far off. About fifty feet away a young bull tore up grass to eat. He came closer and Lee waved her arm at him. "Go away from here," she said.

By observing the Amboseli elephants, Moss hopes to learn more about basic elephant behavior when the animals are not bothered by people. She has learned to identify by sight almost every one of the elephants in the park. Over the years she has kept track of new births and how young elephants grow, as well as those that have died.

Elephant Society

Moss has put together the following picture of elephant society in Amboseli: Elephant families are made up of a handful of young and a few related females, one of which heads the group. Males are driven from the family in their teens. After that they live by themselves or sometimes in the company of other males. They usually join the females only to mate.

The members of a family stay very close together, seldom leaving one another's sides. Each family also keeps company with a small number of other families.

During the yearly dry season, several of these groups of families share a similar area of land and make up a bigger unit called a clan. Some clans stay in the thick bush at the edge of the park, while others gather around the swamp. The bulls generally wander here and there, seldom with the young and females. As the wet season arrives, the elephants leave the park for higher ground outside its boundaries.

When the climate is particularly wet, Moss discovered, these patterns may change somewhat. The rains increase the vegetation and therefore the amount of food available to the elephants. Rather than break up into smaller groups, the elephant families form large herds, including bulls, numbering up to 300 animals.

Moss believes that if the elephants always had their way, they would live in large groups; but when the dry season comes they are forced to spread out because they cannot find enough food for all of them in one place. Whether or not this is true is uncertain, but Moss plans to study the Amboseli elephants for several more years in hopes of learning more about them.

A Day with a Field Biologist

When she is in camp—Moss spends a few days each month in Nairobi editing her newsletter—the American field biologist rises early. After breakfast she gets behind the wheel of her Land-Rover and heads out after the elephants, which can be anywhere in the area. Either sitting behind the wheel or on the roof of her vehicle, she watches the elephants for several hours, taking notes on what she sees. Sometimes she follows them in her vehicle, noting where they are likely to travel in the course of the day.

More than once the vehicle has broken down and Moss, alone in the bush, has had to fix it. In case of real emergency, however, she knows that if she does not return to camp by 7:30 P.M., authorities from the park headquarters, only a short distance away, will set out after her.

I went with Moss to watch her as she observed the elephants. It was a bright, sunny day, warm but not humid. A large group of elephants, perhaps 100, was feeding in the thick greenery at the edge of the swamp only a few hundred yards from the camp. Moss shut off the engine, and gradually the elephants moved until they were feeding all around us.

Moss explained that the group was made up of several families, of about eight elephants each. "There's a family I haven't seen for a while," she said, pointing to several elephants not far off. To me they looked no different from any of the others. Moss, however, can tell them apart with ease. She spent three years learning to recognize the different animals, painstakingly photographing them and studying their pictures.

"There are all sorts of groups here today," said Moss, as she rapidly made notes on paper held in a clipboard. She pointed to where two cows approached one another. Each wrapped her trunk around the other's; the action, Moss explained, was a greeting between two members of the same group.

Moss watched carefully to check on which elephant families and groups of families had gathered together to make up this large herd. "Usually it takes a while to sort out the groups in a big gathering such as this," she told me.

By then the elephants had surrounded us. The air was filled with the sound of their feeding as they tore grass from the ground. Every once in a while an elephant snorted, sending clouds of brown dust into the air. At the edge of the herd one of the creatures trumpeted loudly, an ear-shattering sound, a sound of wild Africa.

As the heat of the day increased, the elephants fanned their bodies with their immense ears. White cattle egrets, birds with long legs and beaks, stalked through the grass around the elephants' feet, picking up insects stirred up by the movement of the big creatures. The egrets were constant companions of the elephants as they fed, sometimes even perching on the great gray backs of the tuskers.

Gradually the elephants moved past us. Moss started the engine and we drove for several hundred yards until we were ahead of the elephants again, still in their slow line of march.

Not far away a gang of small calves played with one another, intertwining trunks and tails, pushing, shoving, and trying to climb on one another's backs. "It looks like a circus," I thought. One of the little elephants was bowled over by its playmates, who quickly piled atop it. In a moment the calf on the bottom wriggled out from beneath the pile, sending the others tumbling. They all seemed to enjoy it immensely.

Finally, it was time for me to leave. A small aircraft was waiting at the Amboseli airstrip, dirt runways with Mount Kilimanjaro in the background, to return me to the city of Nairobi, an hour's flight away. Later, as the aircraft took off and circled Amboseli, I looked down at the green swamp and saw Moss's camp in the late afternoon sun. Here and there scattered through the greenery, I could see the immense gray forms of the elephants that Moss had come to know better than any person in the world.

3

Wildlife Law Enforcement

On January 5, 1981, two Idaho conservation officers—game wardens, as they are commonly called—were gunned down in cold blood by a man they were questioning about illegal trapping and hunting. The wardens were tough, experienced lawmen. They knew the man was a rough customer. He had a reputation for having a mean temper and being a deadly shot. The wardens had disarmed him, and while they asked questions eyed him carefully. Somehow, however, they must have let down their guard, if only for a moment. That was all it took. A hidden gun was drawn, fired; and both wardens fell to the ground. After they fell, they were shot again to make certain they were dead. The man suspected of the killings was later captured after a shoot-out with agents of the Federal Bureau of Investigation. He was tried and convicted of voluntary manslaughter in the case.

This grisly story may be unpleasant to read, but it illustrates something very important about the field of *wildlife law enforcement* that should be considered carefully by anyone interested in entering it. Enforcing conservation laws can be a very dangerous way to earn a living. Many game wardens, for instance, have been killed in the line of duty.

Game wardens regularly work in the wild under isolated conditions. Routinely, they encounter people armed with guns—hunters, trappers, and other outdoorsmen. Most are honest citizens. But some are not. Be that as it may, the odds are good that in the course of his or her career, a game warden will see many, many more people carrying firearms than most big-city police officers.

What Is Wildlife Law Enforcement?

Wildlife law enforcement in the United States today is a broad field. Basically, it is the enforcement of laws for the conservation and protection of wild animals and, increasingly, plants. For most of its history, wildlife law enforcement was concerned with animals that were considered game. Today, because of the awareness that all types of plants and animals are important to the natural scheme of things, the field covers not only game but other species as well. Considerable effort, for instance, goes into arresting people who have violated laws protecting endangered species. More about that later.

The history of wildlife law enforcement in the United States goes back to colonial days. In 1739 Massachusetts communities established the post of deer warden. The warden's job was to make sure that hunters did not kill so many deer that the animals would disappear. Venison, a gourmet dish today, was an important source of food for the colonists.

For most of its history, wildlife law enforcement was poorly organized, carried out catch as catch can. Wardens were sometimes village loafers, given the job of watching out for game because they had nothing else to do. Often a person was appointed warden simply because he liked to hunt or fish. Few were trained in any manner, especially in law enforcement.

By the 1880s, however, a few changes were in the wind. Michigan, Minnesota, and Wisconsin employed the country's first salaried game wardens. Other states soon followed their example. Starting as a local affair, wildlife law enforcement had become the job of the state governments. Each state had a variety of laws covering various animals within its boundaries. Wardens were expected to enforce them.

As the twentieth century dawned, the federal government entered the picture. Congress was enacting laws to conserve wildlife on a national scale. Their enforcement became Washington's responsibility.

Laws involving wildlife deal with many different situations. Some laws specify seasons in which certain game animals may be hunted or how many a hunter may kill. Others ban the hunting of particular species. There are laws that protect endangered species, and those like waterfowl and songbirds that migrate between countries. And there are laws regulating the trade in products made from some wild animals—handbags of crocodile hide and carvings of walrus ivory, for example.

Violation of these and similar laws is called *wildlife crime*. People commit it for many reasons. Sometimes it is just a matter of ignorance. It is not legal, for instance, to hunt sea gulls. Yet a New York State game warden I know once came upon a man with a shotgun and several dead sea gulls on the ground next to him. Asked if he had gunned the birds, the man replied that he certainly had. He was proud of it. He thought that he had legally killed some ducks.

Another reason for wildlife crime is that some people do not take laws about animals seriously. "It's really not hurting anybody else," they reason as an excuse to break the law. These are the kinds of people who will catch more trout than the fishing regulations allow, or catch a songbird and keep it in a cage, although this is not permitted.

When caught, this sort of person often has a very rude awakening. I once witnessed the arrest of a man who was suspected of buying an endangered species of parrot, which had been shipped from one state to another. In doing so, he had broken federal law. Even as he was placed under arrest, he did not seem to realize the deep trouble that he faced. He had just wanted a fancy pet parrot. The seriousness of wildlife crime came home to him, however, when he suddenly was handcuffed like any other criminal, and led off to be fingerprinted.

By far the greatest motive for wildlife crime, however, is greed. There is big money in wildlife. Some people make large sums of money shooting as many deer as they can and then selling the meat on the black market to restaurants. Others smuggle rare birds from far lands into this country to sell for high prices as pets, even though the birds are protected by law. The illegal trade in wild animals and their parts, in fact, is major, big-time crime, a business that earns about $100 million yearly.

Wildlife in the United States is considered public property, not that of the person who owns the land on which it is found. Each state has the legal responsibility for conserving and protecting the wildlife within its borders. Federal wildlife law involves matters affecting more than one state. The federal government, for instance, issues regulations for hunting waterfowl, because these birds migrate seasonally through several states.

Federal laws also cover the shipment of wildlife or products from wild animals between states or in and out of the country. Take the case of a person who wants an endangered species of wild cat as a pet—a very poor idea, by the way. Some states allow it, others do not. As long as the cat remains within the state, however, that is where the responsibility for enforcing the law lies. But suppose the cat's owner decides to sell it to someone in another state. As soon as the animal crosses the state line, the federal government must make sure the law covering it is not broken. In point of fact, the interstate transportation of endangered species is forbidden by federal law except for very special circumstances—by legitimate zoos, say.

The United States has signed many agreements with other nations for the conservation of wildlife. This country has promised, for instance, not to allow the importation of certain rare animals without the approval of the countries from which they come. Making sure these regulations are kept is a federal wildlife law enforcement task.

On a state level, wildlife law enforcement comes under a department concerned with natural resources. Usually the agency has a name, such as Fish and Game Department, Conservation Bureau, or Environmental Protection Agency. The people who ensure that the laws are obeyed are full-fledged peace officers, just like state troopers or highway patrol officers. Some states call them game wardens; others use titles such as rangers, game conservation officers, or wildlife agents.

Several federal agencies play a role in wildlife law enforcement. The Customs Service, for instance, regularly catches people who try to smuggle in protected animals or products. Most concerned with wildlife crime, however, is the United States Fish and Wildlife Service, which has a special division of law enforcement. Working with the division are more than 200 people known as special agents—the same title used by officers of the Federal Bureau of Investigation, Customs, and other federal law enforcement agencies.

At Work with a Warden

The job of a wildlife law enforcement officer, whether for a state or a federal agency, can vary greatly from day to day. Many officers work mainly on patrol—in vehicles, boats, or aircraft, or sometimes on horseback. Even those whose chief task is patrolling, however, may find themselves doing tasks as different as undercover investigation and teaching a safety course for people who are applying for hunting licenses.

Bob Muldoon is a conservation officer for the state of Connecticut. He works in the southeastern part of the state, in an area that is fairly rural, with large tracts of state forest, near the eastern part of Long Island Sound. He is known by outdoorsmen in his area as a lawman who goes by the book, who asks no favors and shows no favoritism. Muldoon works hard to stop wildlife crime such as deer poaching and catching more trout than the limit set by the state Department of Environmental Protection. Like wardens everywhere, however, he has many other responsibilities. One of them is stocking streams with trout from the state hatchery each spring. I accompanied him one day while he worked at this job.

Just after breakfast, Muldoon picked me up in his green patrol car and drove to a meeting point with a truck from the hatchery. Atop the truck was mounted a large metal tank full of water. In the water were 3,000 trout ready to be released into local streams.

Between April and June, Muldoon makes several stocking trips. He knows the streams of his region well because he has spent a long time studying them, looking for places to release trout where they will have a chance to survive. He also must be sure that he puts fish along stretches of stream where anglers are allowed to fish. The trout are paid for by public money. He cannot release them where private landowners forbid others to fish. Each year more and more of the land along the streams Muldoon stocks becomes built up as new homes are constructed. So he must regularly keep an eye on what happens to streamside property.

As Muldoon drove along a country road, followed by the tank truck, I saw another car close behind it. The third vehicle seemed to be tagging along. Sure enough, when we stopped so did it. Out jumped two men, one small and thin, the other big and burly.

"They follow the truck from the hatchery," Muldoon explained about the pair. There are some people, he added, who want so desperately to catch fish, they spend hours waiting for and following the stocking truck. As soon as trout are put into the river, they unlimber their fishing tackle and drop in their lines. They do not seem to care about sportsmanship, just catching fish as fast as they can.

"Maybe they just feel the need to beat the system," said Muldoon. When I walked over to the two men and talked with them, I found this was exactly the case. They claimed that others were somehow—they could not say exactly how—getting more fish from the stocking than they. So they were going to be poor sports and try what amounted to shooting fish in a barrel.

While I talked to the men, the truck driver was scooping trout from the tank with a net. The fish went into a large plastic pail containing water, held by Muldoon. When it was full, he lugged it some 50 feet to the side of a stream. Carefully, he released the fish from the pail. As soon as they entered the quick-moving water, the trout flitted away, vanishing in the ripples. No sooner had the warden released the last one than the two men who had followed trotted to the water's edge and tossed in their lines.

As we headed for the next place requiring stocking, Muldoon talked of his job and how he got it. Now middle-aged, Muldoon was born and raised in the town of Greenwich, across the state on the New York state line. Greenwich today is crowded and built up, but then it had plenty of room for a boy to hunt, fish, and trap. And that, says the warden, is what he loved to do.

"When I was twelve years old," he said, "I'd trap the whole winter." While in the outdoors he kept his eyes open and learned the habits of wildlife. Much of what he learned, according to Muldoon, still helps him today.

After finishing high school, Muldoon joined the United States Navy. He spent four years as a hard-hat diver for the Navy. Although his ambition was to become a game warden, Muldoon found a civilian job as a police officer in Greenwich. He remained in that position for eleven years, but in the meantime took a test for game warden offered by the state. His mark on the test was among the highest of the 500 people who applied. Even so, Muldoon had to wait three years until a slot as game warden was open. When he started his new job, moreover, he had to take a large cut in pay below what he had earned as a police officer with eleven years' experience.

Muldoon says that being a warden is for him the best of jobs. "Even when I'm on vacation, I go to the woods," he comments. Nevertheless, his work can be grueling. When shad—a tasty migratory fish—are heading up the Connecticut River in spring to spawn, Muldoon is often up at three A.M., checking fishermen who catch them commercially to see they are doing it in accordance with state regulations. At almost any time he may prowl the woods at midnight after deer poachers. Several gangs of poachers operate in the area. They kill any deer they can find. Deer can be hunted legally only in certain seasons, with one deer allowed for a hunter's permit. It is not lawful to hunt them at any other time of the year or after dark.

Poachers chase deer year around. During the night they search the forests and fields with large flashlights. When the light strikes a deer's eyes, it is reflected back. The poachers can mark the deer and shoot them. Some poachers play rough. They have fired shots into the homes of people who have reported their activities to game wardens. Yet most of the time Muldoon must work alone. Perhaps he patrols the forest roads or waits hidden in a place where poachers are likely to operate. It takes plenty of self-confidence.

Once he arrests someone, Muldoon has to provide evidence to convict the suspect in court. He has to appear to testify himself. It all takes time.

There is another side to Muldoon's job, too, and it requires the ability to get along with people. When he meets hunters and fishermen out in the field, he must be pleasant, helpful, and, importantly, enlist their support in observing wildlife laws. Wardens know that it is honest hunters or fishermen who are their greatest allies, for they often will report wildlife crime when they see it.

If you are enthusiastic about wildlife conservation, it is a pleasure to see Muldoon or other dedicated game wardens in action. They mean business. I saw a good example one summer day while fishing from my boat on the waters of Long Island Sound.

The boat was anchored near a long stone breakwater. I happened to notice a sleek, white boat edging through the channel between one end of the breakwater and the beach. On board were two uniformed men, Connecticut conservation officers. Once their boat was through the channel, they hit the engine and it streaked off. Suddenly it dawned on me that they were heading to intercept an expensive yacht that was cruising toward land around the seaward end of the breakwater.

Why, I wondered were the wardens after the yacht? On board was just an ordinary family—a man, his wife, and a couple of children. What could they have done to break the law?

The wardens approached the yacht and turned on a loudspeaker. They called for the yacht to stop and let them aboard. The yacht kept going. Echoing across the water came a warden's voice: "This is not a game. You are under arrest."

As the yacht slowed and came about, it turned and I saw the reason the wardens were after it. On one side of the otherwise sparkling craft were large, dripping splashes of mud from the bottom of the Sound, as if those on the boat had been dragging things aboard. For a lark, I imagine, the people on the yacht had been pulling up the pots used by lobstermen to trap their catch and stealing the lobsters found inside them. In lobstering waters this is bad business, a serious crime. Besides, it is a good way to get hurt. Lobstermen work hard at sea. They earn their living from what they catch in their pots. They have been known to shoot pot robbers.

The family on the yacht may have thought stealing lobsters for dinner was a joke, but it was a good bet that from then on they would take wildlife law seriously.

To be a game warden, or at least a good one, you must take your job extremely seriously. The two wardens whose deaths were described at the beginning of this chapter were not the only ones killed in recent years. Two years earlier, for instance, a California warden was fatally wounded on the job. A couple of years before that a deer poacher shot and killed a warden in the back country of Florida. Statistics show, in fact, that a warden has much more chance of being shot and killed than other law enforcement officers. It is a rugged, demanding job, not for the weak of body or spirit. Between 6,000 and 7,000 men and women work at it, and to keep it up they must love what they are doing.

The Federal Special Agent

The morning sun beamed over the woodlands surrounding Quabbin Reservoir in the hills of central Massachusetts. Several men, one a veterinarian from Boston's Franklin Park Zoo, were preparing to release into the wild a bald eagle that had been found injured and nursed back to health at the zoo. Newspaper reporters and television cameramen were there to record the event. A tiny radio transmitter had been attached to the eagle—it was designed to eventually drop off—so its progress after release could be tracked. (The eagle finally headed south, past Chesapeake Bay.)

The television cameras filmed the operation, but the two men who were there to oversee it remained off camera. They were special agents of the Fish and Wildlife Service's Law Enforcement Division. Because the eagle is one of those birds covered by international agreements, the federal government was responsible for seeing to it that the rehabilitation and release of the bird was done according to law. The agents, according to regulations, had to be present when the eagle was freed.

They did not want their faces to appear on television, however, because of another part of their job. Occasionally, the two men worked undercover, trying to get the goods on wildlife violators. The agents, for instance, might pose as unscrupulous bird collectors who wanted to own rare birds even if the law prohibited it. This way they could infiltrate the ranks of smugglers who deal in such birds, which are sold for high but illegal profits.

Releasing an eagle one day, playing an undercover role the next, is very much in character for special agents of Fish and Wildlife nowadays. They often have to be a combination of law officer and naturalist.

Until the late 1960s or so, Fish and Wildlife agents had jobs very similar to those of game wardens but in areas for which the federal government was responsible. Policing waterfowl hunting, for instance, was a major part of their work. In fact, people often referred to the agents as "duck cops." During the waterfowl season they would patrol the marshes, lakes, and seashore, making sure hunters were obeying the rules.

Federal laws set limits for the number and kind of waterfowl that can be killed. Gunners are not allowed to use shotguns capable of holding more than three shells. There are other regulations, too, all of which it was the task of the special agents to enforce.

Fish and Wildlife special agents still work as duck cops. During waterfowl season they regularly slosh about wetlands in hip boots, visiting hunters in their blinds. They cruise creeks and rivers in small boats for the same purpose. Sometimes they fly aircraft patrols. Aerial observation is a good way to spot people who are trying to lure waterfowl by throwing out corn and other feed, which is unlawful.

During the past several years, however, new responsibilities have been added to the agents' job. They have had to become crackerjack investigators, as much so, for instance, as special agents of the Federal Bureau of Investigation or the Treasury Department. The change came about mainly because of two developments. One was the involvement of the federal government in international efforts to save endangered species. Treaties were signed with other countries that had to be enforced. Another was the growth of big-time wildlife crime.

When animals become hard to get or are protected by law, they—or products made from them—can become extremely valuable. Some rare parrots that may not be legally caught and sold bring prices of thousands of dollars each in the black market pet trade. Elephant ivory sometimes is as valuable as gold. There is a legal market for ivory that has been taken in accordance with the law, but there is also one in ivory from elephants that have been poached—illegally killed.

The business of selling black market animals and products became so widespread that in 1979 President Jimmy Carter issued an announcement calling for government law enforcement agencies to declare war on wildlife crime. The key strike force in this war is Fish and Wildlife.

Fish and Wildlife agents brought the war home to wildlife criminals. Within two years of the president's declaration, they carried out two of the biggest operations against wildlife crime ever conducted in the United States. Both depended on careful undercover work.

One was dubbed the "snakescam sting." It was aimed at the huge illegal pet trade in snakes and other reptiles protected by either state or federal regulations. Many people like to keep such creatures as pets. Some of these pet collectors do not realize it when they buy protected animals. Some do not care.

To infiltrate the illegal business in reptiles, two agents went undercover and opened up a "Wildlife Exchange" in Atlanta, Georgia. From the outside it appeared to be a wholesale animal dealership. Really, however, it was a means to gather evidence to put wildlife criminals behind bars. The undercover "animal dealers" bought and sold 10,000 animals that had been illegally captured in the wild, tape recording all their business deals.

Armed with information gained at the Wildlife Exchange, two hundred federal agents and state game wardens picked up search warrants and raided 45 places in fourteen states. They arrested 25 suspects. Among the animals that were illegally traded were many that are dangerous, including water moccasins, rattlesnakes, and Gila monsters, which are venomous lizards.

The other case involved walrus ivory. The walrus is an endangered species and under law cannot be hunted, with one exception: Eskimos, who depend on hunting for much of their food, may kill walruses for their meat. Criminal ivory dealers, however, paid some Eskimos to kill walruses for their ivory tusks. These were then sold as material for carving jewelry and knickknacks.

The walruses were killed in Alaska, but the key to the case was found in New Orleans, Louisiana. There a Fish and Wildlife agent—who was well known for catching people who poached alligators and baited waterfowl with corn in the coastal marshes—was tipped off that black market dealers were looking to sell walrus ivory.

Posing as a wealthy businessman who wanted to buy ivory, the special agent gained the confidence of the dealers. He secretly tape-recorded hundreds of hours of talks with them, during which they offered and sold him tusks. The information he obtained enabled agents to make simultaneous raids in five states from Alaska to New Jersey. They swept through homes and businesses of people suspected of dealing in unlawful walrus ivory. The agents seized 10,000 pounds of ivory worth almost a half million dollars.

Fish and Wildlife agents sometimes team up with special agents from other law enforcement services. Each federal law enforcement department has the responsibility for policing different laws. If a person smuggles a protected animal into the United States, say, for the unlawful pet trade, he or she can be breaking more than one law. First of all, smuggling any item is a crime. Catching smugglers is the job of Customs agents. Fish and Wildlife agents are empowered to go after people who break laws by bringing endangered species into the country without permission.

Customs and Fish and Wildlife agents have arrested many people for smuggling parrots, which bring high prices as pets. Parrots are smuggled for two main reasons. One is that many species are protected. The other is that even with nonendangered species, regulations prevent people from importing birds that carry diseases. Wild birds from exotic places can carry diseases which could spread to farmers' poultry. Some of these diseases can quickly kill millions of dollars worth of poultry. And a few diseases can even be transmitted to humans. Federal law requires that birds imported into the United States be kept in isolation for a month to make sure they are not sick. If the birds turn out to be sick, they are destroyed.

Some people smuggle birds because they do not want to pay the costs of keeping them in isolation. Or else the smuggler is afraid that the birds will be destroyed. Often, moreover, smugglers do not care a bit whether or not the birds they sell are sick. They don't care if the birds die after they are sold.

A great deal of smuggling goes on across the Mexican border, because most parrots sold by the pet business come from Latin America. A Customs agent in Texas managed to infiltrate a group of people he heard were buying and selling smuggled parrots. He discovered that they got the birds from a smuggler who had a business in Laredo, Texas.

The Customs agent and a long-haired, bearded Fish and Wildlife agent posed as parrot buyers to catch the smuggler. (Their names are not used because of their undercover role.) After making contact with the smuggler through an informant, they met him at a warehouse he owned. They said they wanted to buy parrots. The agents had $4,000 in $100 bills as "flash money"—that is, to trick the smuggler into believing they really were going to buy the birds.

The smuggler told the agents the birds were in a rest room. One of the undercover men went into the room and found a batch of parrots in a cage. They squawked.

"Please keep the noise down," said the smuggler.

Then the agents arrested him. They ordered him to place his hands against the wall and searched him for weapons. Moments later in another room they found more parrots, stuffed into mesh sacks, the kind used to hold onions and other vegetables.

"Call my wife and tell her I'm in big trouble," said the smuggler to his secretary as the agents led him away.

Although there are moments of slam-bang excitement in the life of a special agent, much of the work involves endless hours of attention to small details. Consider the case of a bird dealer in Florida who broke the regulations about keeping imported birds in isolation. A special agent named Maureen R. Matthews worked on the case. Over eighteen months she spent hundreds of hours asking questions, checking facts, and helping government lawyers develop the evidence against the dealer.

The investigation was not a cops-and-robbers affair with high excitement. It was a thorough, precise search for the facts that went on day after day, week after week. Matthews helped government lawyers write a warrant enabling agents to search the dealer's business. Such a search warrant must have precise information to convince a judge that the search is necessary to carry out a criminal investigation. If not, the judge will not issue the warrant.

Once the warrant was obtained, Matthews supervised the search. It took thirteen hours of close and orderly inspection. Masses of records were checked and rechecked. Matthews also had to identify and track witnesses, including former employees of the dealer. She had to find experts who could testify about birds and their diseases. Then she had to help the government lawyers put together an indictment, the document under which a suspected criminal is charged with violations.

The dealer was convicted. Government authorities said the conviction was an example that would make other bird dealers think twice before violating the law. Without the dedicated work of Matthews, the authorities said, the case might have gone down the drain.

What It Takes to Get into Law Enforcement

The person who has the best chance of getting into wildlife law enforcement has a background with at least some experience in police work and biology. Of course, such a combination is often hard to get. One way to do it is through education. Most colleges offer courses of study in biology and related fields. Many also have studies in law and police science. Someone who enrolls in such courses has a jump on applicants for wildlife law enforcement jobs who lack this sort of educational background.

You can acquire the needed experience through work, as was the case, for example, with Bob Muldoon, serving as a municipal policeman before becoming a conservation officer. Increasingly, however, state conservation agencies are looking for people who have had at least two years of college in biology or police science—or best, both—to be their game wardens.

It is almost essential that you have a college degree, or even a graduate degree, to join Fish and Wildlife's law enforcement division as a special agent. Some Fish and Wildlife agents even have experience in other federal law enforcement agencies, such as the Federal Bureau of Investigation or Customs Service.

One young special agent, whose name, again, will not be revealed because he sometimes works undercover, explained to me how he joined Fish and Wildlife. Coming from the Boston area, he was a city boy who fished once in a while but never hunted or spent much time in the woods. As a youngster he did not even know that such a job as a Fish and Wildlife agent existed. But he was interested in police work—any kind.

The young man enrolled at Boston's Northeastern University, which has a good reputation for its police science courses. He hoped on graduation to apply to various federal and state agencies, such as the state police. Then a lucky break came his way. A chance arose for him to work while still a student with the office of Fish and Wildlife in Boston. A special intern program for Northeastern police science students had opened up. (Such programs may no longer be available, so do not count on them.) He worked with agents as an assistant, carrying out tasks such as researching laws applicable to various investigations. He did so well that on graduation he was offered a chance to become a full-fledged agent.

After passing an entrance test, new agents of Fish and

Wildlife must undergo extensive training. Along with agents from Customs, the Internal Revenue Service, and other federal agencies, they train at a special government school in Glynco, Georgia, for two months. Study centers on basic criminal investigation. After completing the course, they attend five more weeks of school on Fish and Wildlife's special law enforcement tasks. Then they spend a year as apprentice agents at various posts around the country.

The Meaning of Wildlife Law Enforcement

In today's world it is important to conserve natural resources, including wildlife. Conservation is based on science, understanding how to wisely manage resources for the future. Laws are enacted to make sure that people observe conservation. Unless these laws are obeyed, however, all the science, all the work by legal scholars to develop intelligent regulations, and all the goodwill of people toward nature and natural resources have little meaning. The job of the wildlife law enforcement agent is to make sure the laws are obeyed.

— 4 —

Caring for Wild Animals in Captivity

George Lavigueur walked over to a small, fenced pool at the Mystic Marinelife Aquarium, reached out his hand, and patted the head of a young harbor seal that swam over to him.

"Harbor seals are gentle animals," he said, "but this one is even more gentle than most."

The little seal was one of the most appealing animals I had ever seen. It had a small but plump body of gray spotted with brown. Huge brown eyes, bright and shiny, gave it a truly friendly look. No wonder Lavigueur liked the creature so. Then again, I thought, he certainly has to enjoy animals, because he spends his workdays caring for them.

A tall, blond man, Lavigueur is an aquarium and animal handler—commonly called a keeper—at the aquar-

58

ium in Mystic, Connecticut, operated for the public by the Sea Research Foundation. As a keeper, he belongs to a special group of people who are responsible for the well-being of wild animals in captivity, usually at zoos and aquariums but sometimes also at large nature centers.

Zoos and aquariums keep wild animals so visitors can learn about nature from watching them. Increasingly, moreover, zoos and aquariums are breeding rare animals in captivity so that even if they disappear from the wild the species will not vanish.

The animals in zoos and aquariums are unable to fend for themselves as they would in the wild. So they depend totally on people like Lavigueur. Their lives are in his hands. Keepers such as Lavigueur are responsible for the day-to-day care of their animals. Being a keeper is a more complicated job than many people suspect. Feeding the animals and keeping their quarters clean may be the most important part of a keeper's work, but it is by no means all there is to it.

Does an animal look sick? Has it stopped eating? Are other animals in the exhibit—the correct name for an animal display—bothering it? Or is it fighting with them? Are the exhibit's lighting and ventilation operating properly? Are water drains clear? These are just a few of the things a keeper must keep in mind.

Supervising keepers are generally people called curators. The title sounds a bit fancy, but just think of a curator as the head of a department. Many zoos and aquariums have separate departments for each large group of animals—mammals, birds, or reptiles, for example—in their collections. Each department then has its own staff of keepers. Other institutions have a single department that cares for all the animals (except dolphins and a few others used in performances) and exhibits, which is the case at Mystic.

If you can think of the jobs done by keepers and curators in military terms, you might view the keeper as an enlisted man—a corporal or sergeant, perhaps—and the curator an officer. Both come in contact with the animals, but the keeper works with them on a much more regular basis. The curator needs time for activities such as planning new exhibits and developing breeding programs, and in general seeing to the overall operation of the department.

What It Takes

Curators usually have degrees in the biological sciences. Increasingly, zoos and aquariums insist that their curators have not only undergraduate degrees but graduate study, preferably through a doctorate, behind them. The zoo and aquarium curators I know have come from the ranks of herpetologists (people who study reptiles), mammalogists (who study mammals), ornithologists (who study birds), animal behaviorists, marine biologists, and similar disciplines. Sometimes before they attain the rank of curator, they have served as student curators or else keepers.

Years ago few if any keepers had studied biology.

Many were farm boys used to working with livestock and took readily to handling wild animals. Some aquarium keepers had been commercial fishermen. Today all this is changing. Zoos and aquariums do not demand that people applying for the job of keeper be college graduates, but those with at least some study of biology beyond high school—or, for instance, work experience with animals on a farm—have the best chances of finding work. And more applicants than ever have in fact completed college.

No amount of education, however, can make a person a keeper unless he or she—many women have joined keeper staffs—has a deep and genuine liking for animals and feels a responsibility toward them. By and large this is the reward people seek from the job, which often does not pay particularly well. A keeper, moreover, must not be afraid of hard work—even with modern equipment, enclosures have to be swept and shoveled clean or bales of hay and feed be carried. For a person who enjoys wild animals, however, the job can be unique. Probably no other work brings a person so close to so many wild creatures every day as being a keeper.

Worlds of Their Own

Aquariums and zoos are like worlds unto themselves, with their own populations of animals. Keepers must be very much part of that world to do a good job. The aquarium where George Lavigueur works is located in southeastern Connecticut, near the Rhode Island border, and midway between Boston and New York City.

Because all the animals at the aquarium are the responsibility of a single department, Lavigueur cares for a great variety of creatures, ranging from seals to sea horses. The Mystic aquarium, in fact, houses more than 2,000 water animals from all over the globe. They live both in exhibit tanks and pools. At the entrance to the main building, for example, is a 50,000-gallon pool in which live Steller's sea lions, each weighing up to a ton. Inside the building, visitors to the aquarium walk through halls in which are located 34 tanks of various sizes, housing fish, lobsters, shrimp, and other animals. Each tank—and outdoor exhibit, for that matter—is modeled after a natural setting in which the animals would live if they were wild.

One tank, for example, represents an undersea "forest" of kelp seaweed off the California coast. Through artificial strands of kelp, closely resembling real plants, swim Garibaldi fish, brilliant orange-red. Nearby is a tank with a giant Pacific octopus resting in its cave.

The largest exhibit for fish is a circular, 35,000-gallon tank called "The Open Sea." Swimming through its waters are big fish from the Atlantic Ocean—among them bluefish, striped bass, and sand tiger sharks, some longer than a big man is tall.

The aquarium has a marine theater, a pool of 350,000 gallons where the staff stages regular demonstrations with marine mammals—dolphins, sea lions, and beluga whales, white whales that grow to be about fifteen feet long. The animals do tricks—such as fetching rings thrown into the water—that are based on their normal behavior.

Next to the main aquarium is a series of outdoor exhibits called Seal Island. It covers more than two acres of land. In this area live seals and sea lions from the shores of North America. Each exhibit has a large pool, representing the sea, with rocky beaches and cliffs in the background. The rocks are artificial, made of a material similar to concrete, but look real because they were molded from the real thing and colored as naturally as possible.

The first exhibit a visitor sees is called "The New England Coast." Here the rocks are made to resemble granite, as found on the coasts of northern New England. Live trees and wild flowers grow in crevices and pockets along the rocks. Swimming and splashing in the pool and basking on the rocky shore are families of harbor seals, looking and acting much as they do in nature.

Cliffs of black imitation basalt rock tower over the next exhibit, "The Pribilof Islands." It depicts the coastal islands off Alaska. Living in the exhibit is a colony of northern fur seals, which are native to the islands.

The largest exhibit in the complex represents the Channel Islands off the coast of southern California. The rock simulates brown shale and has ledges for diving from; these are used by the agile California sea lions that live in the exhibit. Sharing the exhibit with the sea lions are northern elephant seals, which get their name from their trunklike snouts.

Life by the Sea

George Lavigueur grew up only a few miles from where the aquarium stands, although in 1957, when he was born, it had not yet been built. It opened in 1973. Lavigueur lived in the community of Noank, on the shores of Long Island Sound. Many fishing boats are berthed there, and most Noank people feel close to the sea.

Like many other boys of the community, Lavigueur learned to skin-dive and scuba dive for fish and lobsters. During high school he enrolled in a course on marine biology and participated in field trips to study nearby sea life.

When it came time for college, however, he enrolled at Southern Connecticut State, about 50 miles west of his home, as a history major. But that soon changed.

"I found out that the school had a very good marine biology program," Lavigueur says, "so I switched to it."

While in college he had the opportunity to take part in marine research. As part of his course he helped the state's Department of Agriculture monitor the conditions of oyster beds off the coast of Connecticut.

Meanwhile, Lavigueur decided to try for a job at the Mystic aquarium, which although open only a few years already had a reputation as one of the best in the country. Nothing was available for him, but he kept trying, and by his junior year in college he found an opening as an attendant.

Attendants assist visitors to the aquarium by answering questions and delivering prepared talks about the animals. Being an attendant at a zoo or aquarium—some places call attendants guides—is an excellent way for a young person to learn about the animals and their needs.

At the beginning of Lavigueur's senior year in college, he finally found the job that he had wanted. The aquarium needed a keeper, to work from Friday through Sunday each week. Lavigueur had no classes on Friday, and since Saturday and Sunday were not school days, he had time for the job. When he graduated from college, the position became full time.

At Work with a Keeper

Depending on his schedule, Lavigueur works either in the main aquarium or on Seal Island. His Seal Island days begin at eight A.M. With another keeper he stops at each of the exhibits and closely observes the animals in them. If the animals all seem to be in good shape, Lavigueur starts his daily routine. His first task is to clean the "beaches" of the exhibits. He sweeps the droppings of the animals into the water, which is cleansed by powerful filters.

When he is on duty in the main building, Lavigueur arrives at seven A.M., so he can finish housekeeping before visitors arrive when the doors open at nine A.M. He makes the rounds of each tank, checking to see that the animals are well and the tanks presentable. If he finds algae growing on the glass, he cleans it. Occasionally a fish may have died and need to be removed.

Every two weeks about ten percent of the water in

each tank must be changed, even though it is filtered continuously. This means that each day the keeper caring for the tanks must partially empty and refill several of them. Since some of the tanks hold thousands of gallons, it is a time-consuming job.

After morning cleanup the animals must be fed. It takes lots of preparation and several different kinds of food to serve meals to all the animals in the aquarium. The seals and sea lions eat fish—mackerel and herring bought from fish dealers and stored in freezers at the aquarium. Seals and sea lions are big, active animals and need a solid dinner. A female California sea lion weighing 250 pounds, for instance, eats a dozen pounds of fish twice daily.

The fish are prepared by keepers in the aquarium's kitchen, a clean room with counters that looks rather like a restaurant kitchen. Seals and sea lions are similar to human diners in that they each have their own likes and dislikes. Some want fish whole. Others will not eat unless the fish has been cut up. There are animals that accept only chunks of fish and others that eat only fish that has been sliced diagonally. Some like fish heads. Others want fish tails. Lavigueur must know the preference of each animal living on Seal Island.

Once the fish is prepared, Lavigueur weighs out appropriate amounts in buckets and carries them to the exhibits. At feeding time, each animal hauls itself out of the water onto the beach and waits to be fed in its own special place—its "station." The animals have been trained to go to their stations or they will not receive fish.

The orderly system of feeding is necessary because otherwise the animals might become unruly. Large residents of the exhibits might push small ones out of the way. Some of them might fight.

"Besides," says Lavigueur, "this way we are sure each animal is hungry. If an animal doesn't eat, it may be sick."

The menu for the fish is more varied than the meals for seals and sea lions. Whipping cream, gelatin, strained vitamins, and even paprika go into the preparation of some of the food served to them.

Basically, according to Lavigueur, there are two main groups of fish in the tanks. The fish from northern waters are generally large species that can manage to eat chopped squid and smelt. Most of the fish from tropical areas are small varieties. Some are given prepared dried fish food, the same kind fed to fish in home aquariums. But others are more demanding.

For these Lavigueur and the other keepers make a special dish. They mix gelatin, whole fish, trout meal, clam juice, water, paprika, and vitamins in a blender, then let it stand until it becomes firm, like desert gelatin. Next they cut it into pieces of various size and feed it to the fish.

Some fish in the aquarium tanks normally live on coral reefs, where they crush the coral so they can eat the small animals inside it. They take a long time to eat. Other fish in the same tanks are fast eaters. They would finish all the food before the coral crushers had a chance to eat very much, except for a trick used by the keepers.

They mix some of the food with plaster of Paris and let it harden into "cookies." The only fish that can break up the cookies are the types that usually feed on coral.

Yet other tropical fish eat vegetable matter—algae—as well as other animals. A heavy growth of algae tends to dirty a tank, so Lavigueur and the other keepers feed these fish vegetables—parsley, broccoli, or celery weighted and dropped to the bottom.

After the regular chores are finished, Lavigueur may be called on to perform any number of other tasks. He may, for instance, help the curator who is in charge of the department arrange coral or rocks in a new exhibit tank. Or else he might help the zoo veterinarian give a seal its yearly physical examination.

A seal is unlikely to sit still while someone examines it. So when it is time to give an animal its physical, the keepers herd it into a cage small enough to keep it from thrashing about. Once the veterinarian finishes, the creature is released.

During the course of a day Lavigueur spends considerable time quietly observing animals, almost like a field biologist in the wild. He keeps regular notes on each one's appearance and behavior. This information goes into the aquarium's permanent records, which are an important tool for keeping animals in captivity.

Each year, for example, seals lose a substantial amount of old skin and hair within a period of a week or so. At this time they lose their appetites. If Lavigueur notices that a seal has stopped eating, he checks the records. If they show that the year before at the same time, the same seal lost its appetite and began to shed skin, the keeper knows the animal is not ill but is behaving normally.

On his daily rounds, Lavigueur watches to see if any of the animals are pregnant or show signs of wounds from fighting. He follows the progress of young to make sure they are growing normally.

Many seals have been born and reared at the Mystic aquarium, which indicates that the conditions on Seal Island are quite healthy. Even in exhibits as large as those at Mystic, however, the mother seals cannot move far from their young when it is time for the young to stop living on milk and begin eating fish. Keepers can encourage most seal pups to change to solid food by quietly giving them fish on a regular basis in a corner of the exhibit away from the other animals. But sometimes this does not work.

In such cases, the keepers may remove the pup from the exhibit, as they had done with the little harbor seal mentioned earlier. Lavigueur spent weeks teaching the friendly little animal to readily accept fish. First he dropped live minnows into the water. The pup immediately snapped them up, acting on its inherited instinct to chase fish. After a while the pup began to eat dead minnows, and finally cut-up herring and mackerel. By that time the pup was ready to go back to the exhibit and take its place in the harbor seal colony.

Occasionally a mother seal refuses to nurse her pup. If this happens in the wild, the pup usually dies. At the aquarium, however, pups whose mothers ignore them are fed milk by the keepers. One keeper holds the pup while the other feeds it through a tube. Later, the keepers change the pup's diet to fish.

Collecting Animals

Lavigueur's duties do not always keep him at the aquarium. Occasionally he goes with other members of his department to collect animals for the exhibits. Mystic aquarium collects some of the animals it needs from nearby waters—the sea, lakes, and streams. Sometimes collecting is a matter of sweeping a large net through a pond or tidal shallows and catching small fish. Other times, however, it means going after creatures as large as sharks and whales.

I have accompanied people from aquariums on many trips to collect animals large and small. All these expeditions were exciting, no matter what kind of creatures we were after.

One cold day in early March, for instance, I traveled with keepers from the New York Aquarium to a river halfway down the coast of New Jersey. Our goal was to collect sea urchins, small, spiny animals that cling to rocks and other hard objects under the water. We put on rubber wet suits and face masks and jumped into the river where it emptied into the Atlantic. Even with wet suits we were chilled to the bone almost immediately. It was difficult work, due to the cold water and currents. Then one of those unexpected things happened that makes working with nature always exciting. As I headed down for another dive of about 20 feet, I suddenly saw several large fish circling round and round in the water. They were striped bass, some perhaps 30 to 40 pounds in weight. We had apparently found a spot where the bass were holing up during the cold weather. I swam into the middle of the circling fish, so close I could have touched them. I remember thinking how happy I would have been to catch bass like those on my fishing rod. The fish paid no attention to me.

Over the years I have made several trips with aquarium teams to capture sharks. One of the most exciting was with Marineland of the Pacific, an aquarium in southern California. We boarded a 20-foot boat one morning and set out into the waters off the coast, where blue sharks—the type we were after—were common. Fog cloaked the water, but soon it broke up and we could see schools of pilot whales surfacing after chasing fish and squid down below.

As we watched the surface of the sea, which was glass calm, we suddenly spotted a triangular fin a few hundred feet away. One member of the team dropped a bag containing chum—a mixture of ground-up fish parts—into the water. The juices from the chum leaked from the bag and attracted the shark toward baited hooks we had set out on hand lines. The blue shark flashed past the boat but failed to take the hooks. Then it dove and vanished.

Before long, however, a second shark appeared and cruised toward us. It moved under the boat and swept past the bait but did not strike. Then it made another pass and this time grabbed the bait. The hook was set in the shark's jaws, so even though it rocketed away the line held. We let the animal swim until it was tired, then hauled it alongside.

Keeping away from the shark's jaws, we worked a canvas sling under the shark and lifted the five-foot-long fish into the boat, where we deposited our catch inside a crate prepared for it. Then we shoved the nozzle of a hose into the shark's jaws. A pump pushed seawater through the hose and out through the nozzle, which had holes along each side. The water flowed back over the shark's gills, providing the oxygen it needed to breathe until we returned to port and put it in a pool at the aquarium.

Hazards of the Job

Not surprisingly, people who work with animals sometimes are injured, although seldom seriously. Lavigueur, for instance, has been bitten by nervous seals and once by a dolphin being moved on a sling similar to the one we used to catch the shark.

Occasionally, however, a keeper or, for that matter, a curator, can have a real brush with danger. One curator I know was bitten by a rattlesnake and spent several painful days in the hospital. A keeper who is a friend of mine narrowly missed being poisoned when a dangerous snake struck at his hand. The animal's fangs straddled his index finger, missing flesh.

Experienced keepers always try to remember they are dealing with wild animals, even though the creatures may act tame. Injuries usually take place when a keeper becomes so accustomed to the job, he or she takes the animals for granted. Working with wild animals, including those you may know well, always calls for a bit of caution. After all, sometimes even pet dogs bite and cats claw. A risk of injury, although not really great, goes with a keeper's job.

Where Are the Jobs?

There are about 200 zoos and aquariums—mostly zoos— in the United States. Some are operated by cities, others by zoological societies, which are private organizations working for the public good and not for profit. Some are commercially operated, as businesses.

Of the total, only about 50 are of good quality, and perhaps half that truly outstanding, like the Bronx Zoo, San Diego Zoo, or Mystic Marinelife Aquarium. It takes a vast amount of money to have a quality zoo or aquarium, and this is not always available to cities or zoological societies.

Many of the poorly run zoos require little in the way of experience or training from their keepers, who sometimes are really little more than handymen. There is no real possibility of careers for dedicated animal handlers in such places. The only good jobs are at the zoos and aquariums that are well run. At the most, however, even a large zoo may have only a few dozen keepers on its staff, so there is not a large job market. For curators, the openings are even fewer. Some places have only one or two curators on the staff.

Openings do occur fairly regularly, however, partly

because the jobs, especially for keepers, do not offer particularly high pay. A keeper cannot expect to earn as much, for example, as most fire or police officers. Even someone as dedicated as George Lavigueur, for instance, was thinking about resigning from his job when I spoke with him about it. Later in life, he said, he hoped to earn more money than was offered for the work he did, as much as he enjoyed it. Lavigueur said, however, that having worked at the aquarium would be extremely helpful to him in the career he hoped to enter. He planned to return to a university for more education, then enter the field of aquaculture, the "farming" of fish and other water animals for food and other resources.

5

Nature Writing

Gasping in the thin air, I slowly climbed a trail toward the top of a ridge 15,000 feet high in the Andes of Peru. The ground underfoot was rough and rocky, and bearded with bunches of tough, brown grass known as *icchu*. The dry stalks of the grass shook stiffly in the winds that seem to blow endlessly across this world above the clouds.

As I reached the top, I stopped for a moment to rest, then looked about. Stretching out on all sides was the high Andean plateau, a rolling grassland higher than many of the peaks in mountains such as the Alps and Rockies. Scattered about the grassland were small groups of golden-brown animals, with slim necks and long legs. They looked like small llamas, and indeed, they were wild cousins of the llama, called vicunas.

I had come to the Andes on assignment for a wildlife and conservation magazine, *Animal Kingdom.* The editor wanted me to write a story about what the Peruvian government was doing to save the vicuna from extinction. Once very common, vicunas were killed in large numbers for their wool, which is the finest in the world. They were nearing extinction when the Peruvian government began to protect them and rebuild their population.

Writing about Nature

The assignment in the Andes was one of many that have taken me to wild and exciting places, where I have gathered information for articles and books on wildlife, conservation, the outdoors, and other topics that generally fall under the subject of "nature."

I enjoy the outdoors and am deeply interested in wildlife. Also, however, I find writing a satisfying way to earn a living. Writing about nature, therefore, has enabled me to combine my most important interests and abilities.

A nature writer does not belong to an organized profession as does, for instance, a biologist, physician, or attorney. Rather, he or she usually is a journalist or other communicator who happens to be covering nature but may, as I sometimes do, also work on other subjects.

Different types of jobs in the field of communications allow people to write about nature, either all or part of the time. Some people work as editors and writers on the staffs of magazines such as *Audubon*, *National Wildlife*, and *Outdoor Life*. A few newspapers have reporters who cover wildlife and conservation, or science writers who sometimes do stories on nature along with technology, medicine, and other affairs.

Many organizations concerned with conservation and nature in general employ writers in their public relations and publicity departments. These writers produce publications that promote the organizations and their goals, or work with the news media toward the same end. A public relations writer for a zoo might write a newsletter that tells about activities and animals there. A writer for a government agency like the Fish and Wildlife Service could write an announcement for newspapers, radio, and television describing a new program to rescue an endangered species.

At this stage of my career I am a "free-lance" writer—that is, one who is self-employed and writes for many different publications and publishers instead of working on the staff of just one. Free lancing, as it is called, allows a writer plenty of personal freedom and a wide variety of assignments. But it can be a difficult way to earn a living, because it does not guarantee a regular salary nor provide benefits such as paid vacations.

Over the years, I have held many of the jobs described in this book. Gradually, in these positions, I tried to write more and more about the natural world, until that is now what I do most. There are many ways to find and qualify for the jobs that involve nature writing. The way I did it is one.

My Way

Looking back, I can really say that things I did as a boy helped me greatly toward my later goal of becoming a nature writer. From my early days in elementary school, I developed a keen interest in animals. I came from the Bronx, in New York City, where the only places with greenery and a few animals were parks. When I was quite young, however, my parents moved to the Connecticut city of Waterbury, which although it was a big industrial community had a few wooded areas, marshes, and ponds, and was not far from rural countryside.

While still in elementary school I began to spend time in the small patches of woodland that were within reach. There I watched for birds and explored small ponds and streams for frogs, salamanders, and insects. There was not an abundance of wildlife in the area, but enough to keep me busy.

At the library I found a book that explained how to keep a notebook, or nature journal, recording what I had seen while observing birds and other animals. I started one, and kept it up. Today, whether in Africa, Asia, South America, or the swamp behind my present home, I still keep a nature journal, recording everything important that I see. I also keep similar journals while fishing and hunting. All of these journals are useful when I need to remember information from the past and details from my research for articles.

If any single activity I carried on as a boy has helped me in my career, it was reading about nature and science. As soon as I was old enough to use the city library, I began borrowing books. At one point I brought a large cardboard box and obtained the librarian's permission to take more books than the limit she generally allowed. Reading regularly about wildlife, especially, created a background of knowledge that has remained with me and that I have fallen back upon time and time again as I have written my assignments.

So captivated did I become by animals that I decided I wanted to study them as a scientist when I became an adult. Of course, I never did. There are several reasons, but to make a long story short, it came down to a matter of details. A good marine biologist or field biologist—like Cynthia Moss, for instance—has to pore over details. One famous zoologist I knew once spent weeks noting down the time a single turtle took to sun itself day after day, all the while keeping track of the air temperature and even the position in which the creature rested. I like turtles and have fun watching them, but that sort of detail, absolutely important to the scientist, is not for me.

Many scientists, moreover, must confine themselves to a very small area of study. Another biologist I knew spent years studying only the behavior of fiddler crabs. Still another was concerned solely with whales. My interests are too broad for this sort of specialization. Nature writing takes a wide knowledge of natural history and science, but does not demand great expertness in details, although to be sure that does not hurt.

As for so many other young people, high school

brought a time of confusion for me. I became uncertain about what I wanted to do in the future. Although I discarded the idea of becoming a scientist who worked with animals, at least they continued to occupy my interest. By the time college arrived, I decided that my abilities suited me more for communications than any other field, so that is what I studied.

After college I worked for various newspapers in the New York City area, beginning as an apprentice and graduating to full-fledged reporter. I became skilled at handling assignments that brought me considerable action. I covered gangland slayings and the trial of a man charged with spying against the United States for the Soviet Union, and mixed with patients inside a state hospital for the mentally ill to discover if they were being mistreated by the staff. On the scene at a riot in a large city, I had to dodge sniper bullets. I suffered frostbite covering a major fire in below-zero temperatures. At 23 years of age I posed as a teenager to determine if young people's drinking alcohol in taverns that catered to them was contributing to highway accidents. All of this experience was to eventually help me in nature writing, although I had no inkling of that at the time.

A New Assignment

One day the opportunity arrived to cover a new type of story. It involved what is called "industrial espionage," the stealing of secrets from one company so another one can profit from the pirated knowledge. Many types of industrial espionage are against the law, because legally the companies who are victims own the secrets just as they might machinery or laboratory equipment.

Industrial pirates from Europe were infiltrating American pharmaceutical companies and stealing secret processes used to make antibiotic drugs. The processes were worth large amounts of money, because they made the job of manufacturing expensive drugs easier. To cover the assignment I had to not only call on my experience as an investigative reporter but learn something about the science and technology of the antibiotic industry. All of a sudden I had an assignment that brought me into contact with one of my old interests, science, and I liked it.

A little at a time, I pressed my editors to let me cover more science stories—new developments in laboratories, medical discoveries, and advances in the then-new American space program. I found I liked science reporting better than any other kind.

To progress as a science reporter, I needed more education in the sciences. I found a program at Columbia University that provided fellowships for reporters who wished to study science there for a year. Graduates of the program usually developed into science writers whose work was respected and in the public eye. It took two years before I was accepted as a fellow at Columbia, but one September day I found myself there ready to begin studies.

At Columbia I was able to study a broad range of sciences but concentrated on those related to biology. While at the university, moreover, I found a valuable part-time job. It was as a writer in the public relations department of The American Museum of Natural History, in New York City like Columbia. The job involved writing press releases—stories about museum exhibits and programs that were sent to the news media. For the first time, I had the chance to write about animals—as well as about other aspects of natural history. It was an important turning point and started me thinking about nature writing as a career.

Magazine Editing

After graduating from my fellowship program, I became an editor in the science division of Scholastic Magazines, which published science magazines for students in junior and senior high school. As before, I wrote about various types of science, including, every so often, something on nature. While there I was given my first assignments involving travel to places that then seemed very far away. I went to Puerto Rico to write about an astronomical observatory there, and to California, where I reported on how the Marineland of the Pacific aquarium caught sharks for their exhibits.

Writing regularly about science increased my understanding of it and the way in which scientists go about their jobs. At the same time, I learned what it took to publish a magazine regularly.

The New York Zoological Society

Both skills helped me when I applied for a job that had opened at the New York Zoological Society, which operates the Bronx Zoo, New York Aquarium, and Osborn Laboratories of Marine Sciences. The position was as curator of publications. The curator headed the department that published the society's magazine for members, *Animal Kingdom*; a scientific journal; and numerous pamphlets and reports issued from time to time. The department was also responsible for public relations, as well as for publicity aimed at encouraging people to visit the zoo and aquarium. My total experience—newspapers, public relations, magazine editing—together with an understanding of science got me the job.

As editor of the magazine, my job was to find people who could write well about animals, purchase their work, locate and purchase photographs, and see that the magazine was printed on time. Meanwhile, along with my small staff, I had to issue press releases about new animals or exhibits, the society's conservation projects, and research findings by its biologists. If a television crew or newspaper reporter wanted to do a story on the aquarium or zoo, we made the arrangements for coverage, making sure that staff members were around to answer questions and that there was a chance to film or take photographs.

Another part of my job was to visit schools, clubs, homes for senior citizens, and similar places to talk about animals and publicize the zoo and aquarium. Often I would take small tame animals with me to show my listeners. As time went on I began to appear with animals on television programs, serving the society by promoting understanding of wildlife and also attendance at the zoo and aquarium. Zoos and aquariums generally charge visitors an entrance fee to help pay for their operations.

Whenever I could, I found time to be with the staff members who worked with animals. I needed to learn all I could about the society and its activities. So when the chance came to go on an expedition—such as one to capture small white whales called belugas in Hudson Bay —I went along and helped. If the keepers at the reptile house wanted an extra hand roping and tying down a crocodile so a veterinarian could examine it, I was ready. When it was time to escort a new killer whale into its pool, I put on a wet suit and helped guide the animal. To learn more about what the society's biologists were doing in the field, I visited its research station in the jungle of Trinidad and explored caves where the scientists had studied the behavior of bats. All the while my knowledge about wild animals increased.

Although it has grown into a major publication, the society's magazine was then quite small. So while I was a magazine editor myself, I was not well known to the editors of most large publications. Even so, however, I began to sell articles written after my regular workday to other small magazines, which could not afford to pay the fees asked by nature writers at the top of the field.

Eventually, however, being a curator of the zoological society paid off. A publisher asked me if I might like to write a book for young people about animals. And I had an idea—a story for very young children about catching beluga whales and keeping them in aquariums. The idea, and later the manuscript for the book, were accepted, and suddenly I had become an author.

With one book finished, an offer came to do another one. I knew that some writers are able to earn their living working as free lances, and after thought I decided to try for myself. I saved money from my job, began to discuss additional books with publishers, and tried as hard as I could to think of article ideas for major magazines.

Meanwhile, as a representative of the society, I had been appearing regularly on a television program for young people in New York City. It was called *Patchwork Family*, aired over WCBS-TV. I was the "Animal Man," who appeared with various animals and told about their habits. Hearing that I was leaving the society, the producer of the program asked me to continue on the program for a fee, which would help provide a living.

The Free-Lance Life

Leaving the society was difficult, but I had to see whether I could manage as a free-lance nature writer. The first real task facing me was to publish an article in a well-known magazine. I had just visited Puerto Rico, and had spent time in the mountains there observing the rare Puerto Rican parrot and talking to the scientist—Cameron Kepler, mentioned in Chapter Two—who was trying to save it from extinction. Would a big magazine be interested in an article about the bird? I tried *Audubon*, the magazine of the National Audubon Society and one of the world's most respected nature publications. My hopes were not high, but to my surprise the answer was yes. I wrote the article, it was accepted, and I have been writing for *Audubon* ever since. Most importantly, selling an article to *Audubon* gave me a reputation as a nature writer and made me feel much more confident that I could work as a free lance.

With time I began to write for other magazines as well. I published additional books. My association with the zoological society, moreover, had not entirely ended. From time to time I worked as a consultant for the society, helping out with public relations. The society's magazine had started to grow into a publication of increasing stature, too. I began to write for it as well, and writing for one's old magazine can give any ex-editor great satisfaction.

At this point I had an important decision to make. If I was to continue to improve as a nature writer, I needed to see wildlife in many parts of the world. I had no choice but to finance my own trips to Europe and Africa. The information I gathered there helped me write more books and articles, but the trips were terribly costly. Since then I have been luckier, in that magazines and organizations have paid me to travel on assignments.

After about five years as a free lance, I found a new approach to nature writing. Why not combine my old newspaper experience as an investigative reporter with knowledge of natural history? Around the world people were engaged in plenty of questionable activities that affected animals and therefore needed investigation.

In southeast Asia, for example, immense numbers of birds were being caught, taken from the wild, and sold to the pet business. Often they ended up as pets far from their native habitats, in the homes of Americans and Europeans. Some of the birds were rare and protected by law. They were, some conservationists said, being smuggled. Others were common, but captured in such large numbers conservationists worried they would be endangered.

What really was happening? *Audubon* magazine sent me to find out. I visited outdoor markets in places such as Hong Kong, Singapore, and Bangkok, seeing what birds were being sold and in what quantities. I talked to dealers, sometimes posing as a buyer of rare animals. While this time the subject was wildlife, the techniques I used were the same ones I had used as an investigative newspaper reporter years before. It was also similar to being an investigator for a wildlife law enforcement agency, only my aim was not to arrest people but to write about what they did.

Again posing as someone who wanted to buy unusual birds and bring them home to the United States, I gained entrance to compounds owned by people suspected of smuggling animals, along with other things. Some of these people were tough, even cutthroats, but no worse than the criminals I had interviewed while a newspaperman.

Sometimes on assignment I have encountered events that seemed almost as if they were from a movie or television program instead of real life. While in Thailand, for instance, I managed to enter the headquarters of a suspected animal smuggler who turned out to be a strikingly beautiful woman. She was also quite unusual. As one of her servants poured me a cold drink, a huge Great Dane, with a collar of spikes around its neck, eyed me unnervingly. As if that had not been enough, another of her pets, on a long neck chain, sat in a corner. It was a huge leopard, which from time to time she patted on the head. Worse yet, coiled on a cushion in another corner was still another pet—a monstrous python, eighteen feet of snake thicker than a man's leg. I seldom have been happier to leave any place than that room in Thailand.

My old magazine, *Animal Kingdom*, has encouraged me to take assignments that require an understanding of wildlife and conservation as well as politics, economics, and social problems—things almost every newspaper reporter learns about.

One such story involved the vicunas of Peru's Andes. Scientists working in behalf of the Peruvian government had brought the vicunas there back from the edge of extinction. One of the methods they used was to convince Indians living in the Andes to allow vicunas to graze on their lands, along with domestic llamas and cattle. The Indians opened the lands to the vicunas and they prospered.

The government, however, had promised the Indians that once the vicunas were numerous again, some could be hunted for their wool, which brings high prices. Some of the profits would be given to the Indians, who are desperately poor.

Conservationists in and outside Peru began a heated debate over whether or not it was right to hunt even small numbers of vicunas. One group said that a low number killed would not endanger the species, and that if the wool were not sold the Indians might want the vicunas taken off their land. The other said that hunting would imperil the species again, because they saw no way to control the hunting. Peruvian politicians took different sides on the issue, and the Indians grew angry because they had received no reward for the use of their pastures. So I went to Peru, looked into the controversy, then at home again wrote about it.

In my article I attempted to present both sides of the argument, detailing the reasons offered by each to support their points of view. At the same time, I described in depth the natural history of the vicuna so readers could understand the habits and needs of the animal that was the center of the arguments. I also reported the views of the Indians about the whole issue, which had confused them greatly. By publishing my article, the zoological society's magazine hoped to clarify the controversy so people interested in the future of the vicuna could make up their minds on what course to take.

Some Pointers

A writer's organization to which I belong asked me to provide some hints for success in free-lance nature writing. After I thought about it, some of the advice I had to offer surprised even me.

My type of nature writing, as mentioned, demands a skill for investigation as well as natural history. You need a solid background in subjects like zoology and wildlife conservation—so you will understand not only animals themselves but what scientists who work with them are doing.

There are, however, some qualities and abilities that are also important but perhaps more difficult to attain. You must be able to work under strange conditions in places that are brand-new to you. Many times, for instance, I have arrived on assignment in a country I have never visited before, in the middle of the night, not understanding the language, occasionally even unsure of a place to stay. Unlike a reporter on the staff of a network, news service, or news magazine, a free lance seldom has assistance, much less access to an office, in a strange country.

You should appreciate people of various cultures and get along with them. Often they can be of great help in finding information. I once befriended a young man of the Masai tribe, an East African group famed as warriors and lion hunters. He had left his village and attended school, learning English. After I did him a favor, he helped me back. I wanted to know about how the Masai hunted lions for a book I was writing. My young friend introduced me to some warriors and served as an interpreter so we could understand one another.

Having a strong stomach is important if you are going to cover stories far afield. Strange food and, for that matter, difficult living conditions must not discourage you. You also should be in good physical shape, and have outdoor skills—be able to climb a ridge in high mountains, rope down a sea cliff, or perhaps dive with scuba. All of these things have been helpful to me.

If you are going to write about wild animals, more over, you must feel comfortable with them. Respect for them is one thing, but fear of wildlife is another. I would not go out of my way, for instance, to provoke an African buffalo, one of the toughest animals anywhere and considered dangerous big game. Yet one night on the African plains I returned late to my tent and noticed something moving in the darkness under the awning that extended from one side of it. I assumed it was a zebra, since several of them were grazing nearby. I peered into the darkness and ended up nose to nose with a big buffalo. It looked at me and blinked. If I had panicked, the creature might have become frightened and injured me. As it was, I quietly backed away, went inside the tent, and left it alone. It did the same to me.

Other Ways

As mentioned earlier in this chapter, my route is only one way to becoming a nature writer. I know others who have done things differently, although they have almost always been interested in nature most of their lives. One young man studied journalism at a major university and graduated with honors. For a short time he worked in public relations for a school system in a small town. He was not satisfied, however, because he wanted to do something involving wildlife and conservation. So he quit and took a job at a large zoo as a bird keeper.

His reasoning was that after serving as a keeper, he could progress to a position of somewhat higher rank, perhaps as an assistant curator. Other people with similar aims, however, had degrees in biology and were therefore better qualified for the job. He was in a fix, because although he liked tending birds, he did not want to stay as a keeper. Some people enjoy it for life, but he needed another type of position to be happy.

I met him while I was editor of *Animal Kingdom.* Since he had studied journalism and had learned about animals by taking care of them, I suggested, why not try writing about them? He tried an article for the magazine. Although his writing was a bit rusty, he worked hard and it was published. Nature writing, he concluded, was not a bad way to work at all. After a little looking, he used his combination of talents to land a job as one of the editors of a well-known natural history magazine. The last time I spoke with him he had just returned from a trip to Africa.

Gary Soucie is executive editor of *Audubon* magazine. His first writing job was producing employees' manuals for a bank in California. Soucie, like me, was a young naturalist. He caught frogs and lizards, fished, and hunted in his native Illinois. In college, however, he studied literature.

After graduation he joined the California bank, but before long left for a small newspaper on the windswept and beautiful Outer Banks of North Carolina. Soucie is an avid fisherman, and the Banks are one of the best places to fish salt water anywhere.

Soucie's goal, even though he liked the Outer Banks, was to work for a book publisher. So he moved to New York City, but he had difficulty in finding the type of position he wanted. He took a job writing public relations for an international airline, not because he particularly enjoyed it but for the pay.

While walking one day in New York City, Soucie noticed that the air was filled with pollution—"coal dust," he remembers. "Nobody either knew it or cared," he says. So he decided then and there to work for a conservation organization dedicated to improving the environment. His next job was as the east coast representative of the Sierra Club, which is based in California. After that he took over as head of another conservation group, the Friends of the Earth.

In these jobs, which required some public relations writing, Soucie became aware that there were many different developments affecting the environment that needed to be explored. The way to do it, he concluded, was by magazine articles. Like me he began to write for *Audubon*, and soon he became a full-time free lance. So much did *Audubon* like his work that he was made a field editor, still a free lance but more closely connected to the magazine than others. After eight years, however, Soucie decided free-lancing was no longer for him. Because a free lance does not receive a regular salary, it is sometimes difficult to pay the household bills. *Audubon*'s chief editor had offered Soucie a permanent job several times. When he made another offer, Soucie accepted it.

As executive editor, he develops ideas for stories, assigns them to authors, edits them, and overall helps keep the magazine running. He also has the chance to write articles for *Audubon* himself, and sometimes for other magazines, and he writes books. Much of his writing outside of *Audubon* is about something he seems to enjoy more than any other outdoor activity—fishing.

6

Marine Biology

The sheer, towering cliffs of Tatoosh Island loomed above the mist that cloaked the sea where the Juan de Fuca Strait meets the Pacific Ocean. I stood on the bobbing deck of a salmon-fishing boat that I had boarded a half hour before at Neah Bay, a small fishing village within the Macah Indian Reservation at the tip of Washington's Olympic Peninsula. Once the stronghold of the Macah Tribe, Tatoosh is a huge chunk of rock, sheer on the sides and flat on top. It now is the site of an automated Coast Guard lighthouse.

As I surveyed Tatoosh from a distance of about a quarter mile, I noticed a small rubber boat, powered by an outboard motor, bouncing toward me through the waves. At the controls was Dr. Robert T. Paine, a scientist from the University of Washington, who carries out research on Tatoosh. His specialty is the study of plants and animals that live along the edge of the sea, in the area swept by the tides—the intertidal zone.

Paine had come to take me to Tatoosh, where for a week I was to observe his work. The salmon boat could not make landfall in the booming surf that smashed against the island, so I had to transfer to Paine's small craft. He guided it alongside; I tossed in my gear and then jumped after it. Paine gunned the motor, and we slid across the swells toward the island.

Paine pushed the boat through masses of giant seaweed called kelp, growing just offshore. The plants tossed to and fro in the churning waves. We were headed toward a small, gravelly beach at the foot of a trail that led up the cliffs to the top of the island.

The boat shot through the surf and skidded onto the shore. Quickly we jumped out, waded ashore, and dragged the boat up on the beach. Then, shouldering my gear, we headed for the top. Paine had a camp in an open, abandoned shed not far from the lighthouse. There he and his students who accompanied him to the island lived while studying the life of the intertidal zone.

Because it is sometimes underwater and sometimes exposed to the air, the intertidal is a harsh place for life to exist. The plants and animals living there must in effect adapt to two worlds, of land and sea. This places great demands on intertidal life. Due to these traits, the intertidal is an excellent place for scientists to observe how animals relate to one another and cope with their environment—in other words, ecology. Paine is by his own words an ecologist who uses the intertidal as a living laboratory. Because the organisms he studies are by and large marine, however, Paine's work falls into the broad and exciting field of *marine biology.*

What Is Marine Biology?

In a sense there is no such thing as a "marine biologist." Marine biology is a term that is used quite loosely. Basically, it is any sort of research by a biologist involving plants and animals of the sea.

The biologist may be a specialist in fish, invertebrates, algae, or any of several other areas. He or she may, like Paine, be an ecologist, interested not so much in specific plants or animals but how they interact with one another and their surroundings.

A biologist may study sea life purely for the sake of learning more about it or else to help its conservation. On the other hand, a project in what is called marine biology may have nothing to do with learning about life in the ocean. Instead, it may be the study of a form of ocean life to reveal more about a biological principle. A biologist I have known for many years, for example, has carried out a vast amount of research on how sea urchin embryos develop in the egg. He is not so much interested in sea urchins but in the development of embryos generally. The sea urchin egg just happens to be suited to laboratory study. Even so, this scientist can be considered to be involved in marine biology.

Biologists like my friend spend most of their time in a

laboratory. Other biologists work outdoors, much as field biologists do on land, except the fieldwork is in or on the water. Such people may spend long weeks, even months, aboard research ships. It can be an exciting life but also one that requires constant attention to details. Life at sea can sometimes be downright boring. It also can be harsh.

The only time in my life that I have been seasick, for instance, occurred in the mid 1960s while I was working aboard an oceanic research vessel. Our job was to dredge up samples of shells from the bottom for days on end. The vessel, working off the southeastern United States, would begin each leg of the voyage by steaming to a point 50 miles offshore, then heading directly toward land. Along the way we lowered a dredge to the bottom and periodically brought up samples.

As it turned out, a hurricane developed to the north of where the vessel was sailing. For a week we rode out the storm at its edge, continuing to work. As the storm moved, it pushed us steadily south. Huge swells, often towering over the ship, rocked us. I became violently ill and could not keep anything in my stomach. Part of the time I lay suffering in my bunk, the rest I tried to help with the dredge, sometimes in driving rain brought by the storm. By the time I recovered, I had lost eight pounds.

Some scientists engaged in marine biology use scuba gear to dive and observe sea creatures in a natural state. They must be expert at diving techniques and in good condition, particularly if they operate at great depth, or in seas that are rough and cold.

A major advance for biologists who work under the surface has been the development of undersea chambers in which they can live and work just as they would on the surface. A few years ago I accompanied a biologist who was observing coral reef fish from one of these chambers, resting on the bottom 50 feet below the surface in the Bahama Islands.

The chamber looked like a giant tin can with windows turned on its side and standing on four legs. It was filled with air pumped from a platform floating above. Inside were chairs, tables, counters, and bunks for researchers to rest.

To reach the chamber we donned scuba gear and dove from the platform. When we reached bottom, we swam beneath the chamber, shed our scuba gear, and simply climbed in through an open hatch. Although the hatch was wide open, water could not enter because the pressure of the air inside the chamber was slightly higher than that of the sea outside.

Sitting comfortably in the chamber, we could watch the fish for hours, all the while taking notes about our observations. Undersea chambers are in effect laboratories that enable researchers to make observations easily over long periods, and to conduct experiments in the habitats of creatures that live below the surface.

A Changeable Realm

Similarly, Robert Paine probes the activity of life where the sea meets the land. He has centered his fieldwork at

Tatoosh Island because it has an especially rich variety of intertidal life that, because the island is isolated, is undisturbed by human activities.

The range of the tides at Tatoosh, moreover, is great. Normally there are a dozen feet between high and low tides. A few times yearly, the span from the water's edge at ebb tide to the touch of the spray at high is 20 feet. Thus, a broad band of the intertidal is revealed to Paine and the student researchers who usually accompany him.

It was during one of these series of great tides that I accompanied Paine to Tatoosh. Because of its immense cliffs and boulders, Tatoosh is a spectacular place. But the colors and shapes of its intertidal life almost make one forget the scenery.

The intertidal of Tatoosh teems with plants and animals. The kelp that grows along the lower edge of the zone may reach 50 feet long. Sea stars of red and purple are big enough to cover dinner plates. Cup corals the size of a dime are electric orange. Giant barnacles as big as softballs encrust the rocks. Crevices are the homes of sea anemones, animals with long, delicate tentacles that resemble flowers.

Intertidal life, however, is anything but delicate. Few forms of life live in such harsh surroundings. Waves smash down with a force that in the long run wears away rock. For hours a day the animals and plants of the intertidal must function as sea creatures, below the surface. The rest of the day they are exposed to the air. One moment they are under salt water. The next they are soaked by the fresh water of rain. When covered by the sea, they are largely protected from sharp changes in temperature. When the tide is out they may bake or freeze, depending on the season.

The key to survival in the intertidal is adaptation to these difficult surroundings. Increasingly, scientists like Paine are finding out more about how intertidal life manages to do it.

A Climb to the Bottom of the Sea

Paine and his students visit Tatoosh many times a year. They usually stay for several days. The shed in which they camp was a plumbing shop when the lighthouse was staffed by a Coast Guard contingent several years ago. In one corner of the shed Paine stores groceries brought from the mainland. In another he and his students place their sleeping bags for the night.

They cook some of their meals on a small gas stove, but supper is made over the coals, in a fireplace Paine built of cinder blocks in front of the shed. Occasionally the researchers catch sea bass on rod and reel and cook them over the fire to make a special treat.

Paine's day is timed according to when the tide ebbs, which often means he must be up and about before the sun rises. He walks across the top of the island through head-high brush, largely salmonberry and cow parsnips. After descending the trail to the foot of the cliffs, he sets out for various sites along the shore where he has been carrying on observations.

When the tide recedes from Tatoosh, it seems as if a

new world has been revealed. The landscape is strange and rugged. Huge boulders separated by channels in the rock loom where hours before the waves hid the bottom. Sea caves in the cliffs yawn darkly. Often the entire scene is masked by thick fog.

Clambering over slick, wet boulders—some as big as houses—takes a bit of experience. Paine and his assistants travel over these rocks as nimbly as mountain goats, but newcomers often find the going tough. How tough, I found for myself when I roped down a cliff to see a portion of the intertidal uncovered only during the lowest ebb tides.

From the brink of the cliff, a descent of more than 100 feet awaited. A rope was anchored into the ground at the top by a large iron spike. Holding the rope, I carefully worked my way down a series of natural steps in the rock. The "staircase" ended at a narrow trail that crossed the cliff face.

As I made my way along the trail, I had to avoid a number of glacous-winged gull nests with their brown and speckled eggs. Angrily the adult gulls wheeled overhead, screaming and spattering their droppings in protest. Along the edge of the cliff birds called tufted puffins, with bristly plumes sprouting from behind their heads, peered out of their nesting burrows. On another ledge pigeon guillemots, slate-colored birds with white wings, uttered shrill cries, revealing bright-red mouths.

The trail led to a lip of rock over which was draped another rope. Hanging on to the rope, I slowly went over the edge, to land feet first on a huge boulder. Down below, water surged through a deep crevice in the rock. Beyond that, waves blasted against the foot of the cliffs. From this point I crept down through a jumble of boulders and finally dropped onto a broad tongue of rock, exposed to the air only a few times a year.

I felt as if I had climbed to the bottom of the sea. Behind me cliffs rose straight up from the dripping rocks on which I stood. Underfoot was a slick, squishy mat of sponge covered with brown kelp seaweed. I was actually standing on a bed of life. The sponge was honeycombed with channels and holes, through which it filtered tiny bits of food from the water. In these openings thousands upon thousands of small animals lived. They went about their daily lives in and on the sponge, eating, fighting, breeding, and sometimes dying.

Exploring the inside of the sponge was like touring a miniature underwater zoo. There were sea worms no thicker than a thread, kelp crabs that could fit on your thumbnail, barnacles, and sea stars of all sizes. Tucked here and there in the yellow mat of sponge I saw soft little animals called tunicates, which looked like orange dots encased in jelly. The tunicates looked comfortable, snuggled in the sheltering sponge. Actually, however, the sponge and tunicates were competing for living space—in reality trying to destroy one another to gain more room.

Although you may not always detect it, there is continual competition between living things in the littoral—the edge of the sea. The prize for the winners is continued life, the penalty for losing often death. Mussel beds overgrow the rocks and cover barnacles, smothering them. Plants called sea palms get footholds among the mussels and push them out. Big snails eat barnacles and barnacles crowd sea palms. A beautiful splash of purple on a rock is mixed with one of red. They are really simple animals called hydrocorals in a slow-motion battle for territory.

Studying a Frozen War

Over much of the intertidal, the war for space and food goes on so slowly it seems frozen to the human eye. Increasingly, however, scientists like Paine are putting together a picture of how intertidal life competes and how that affects the way plants and animals are distributed around the zone. For many years scientists believed that the life of the intertidal was arranged in easy-to-recognize belts, parallel to the shore. The life that could cope with exposure to the air least was farthest down in the zone. This is true to a large degree, but is not the whole story, according to what Paine and others have learned.

As an example, Paine pointed to the mussel beds that covered much of the rock in the middle of the Tatoosh intertidal. The mussels can grow no *higher* because they cannot stand longer exposure to the air. But what keeps the shellfish from advancing *lower* in the zone? Careful research by Paine revealed at least one of the reasons.

Clinging to the rock faces below the mussel beds are hordes of sea stars. These creatures use the suckers on their arms to pry open the shells of the mussels, and then eat the soft bodies of the animals inside. Paine kept the rock faces below portions of the mussel beds free of just one kind of sea star known to feed on the shellfish. Eventually the mussels advanced lower into the zone. When the sea stars are present, however, the shellfish are eaten back. Why don't the sea stars advance higher into the mussel beds? They cannot cope with longer exposure to air. Thus, Paine explained to me, a series of forces influences where different types of life congregate in the intertidal.

What You Need for Practicing Marine Biology

Few people get anywhere in the field of marine biology without graduate degrees. A master's helps, but better yet is a doctorate. Paine, for instance, graduated from Harvard with a bachelor's degree in biology, then from the University of Michigan, in 1961, with a doctorate in zoology.

Not until after Paine entered graduate school did his interest turn toward research involving marine life. In fact, between his boyhood and graduate school several different facets of biological science appealed to him, none of which really related to plants and animals of the sea. Even so, he says, all of his varied scientific interests helped shape him for his career.

Paine grew up in the Boston suburb of Cambridge, site of Harvard University. His was an old New England family, with a long tradition of association with scholarly endeavors. His father was curator of Oriental arts at the Boston Museum of Fine Arts but was also an expert mycologist—an authority on mushrooms. For generations, Paine tells me, people in his family had demonstrated a deep interest in nature. So, he says, when as a young boy he brought home salamanders, birds, and other creatures to study, his parents encouraged him. His mother never minded when he showed up with new animals at home.

Birds were Paine's first love as far as natural history was concerned. He watched them as he explored the woods for mushrooms with his parents. By the time he was twelve years old, he had met several prominent ornithologists—scientists who study birds—living in the Cambridge area, which because of Harvard has a large community of scholars.

From them Paine derived a thorough background in birding. After he entered Harvard to study biology, he kept up his associations with the older bird experts.

"By the time I graduated from college, I had eight years of intensive training as a birder," he says.

Paine stresses that even though he did not follow ornithology as a career, it was the training as an observer of natural history that was important.

The Naturalist's Eye

Birding, says Paine, helped him develop a naturalist's eye for looking at the world. He says, "It has served me excellently many times. It provided me with the interest in the natural world, and the curiosity to look for what might be there that doesn't appear at first."

Paine eventually became interested in invertebrates, animals without backbones, because "they enabled me to ask more questions." Because invertebrates are less complex creatures than birds—or even fish, amphibians, and reptiles—scientists can sometimes use them to study nature's interrelationships at a much more basic level than higher animals allow.

The invertebrates that began to interest Paine as he completed college, however, were not living ones but those of the past. At that time several highly respected paleontologists—scientists who study fossils—were on the Harvard faculty. Some had made astonishing discoveries, including those of tiny fossil organisms older than any previously known.

Paine decided he would pursue his interest in prehistoric invertebrate life in graduate school, which he entered after a stint in the United States Army. At Michigan studies in paleontology were under the geology department, so that was the one in which Paine enrolled.

The invertebrates about which Paine decided to conduct the research that would earn him a doctorate were fossil brachiopods. These animals, of a type still existing today, are small shelled creatures that live only in the sea. They were extremely abundant in prehistoric times, so their fossils provide a thorough record from the past.

One of his professors, however, encouraged Paine to study living brachiopods instead of fossil varieties. He accepted the idea but, since the subject was alive, had to carry out his research under the zoology department. Therefore he changed his field.

Paine's doctoral research took him to the shores of Florida, where he spent several months observing brachiopods living there. "It got me into marine biology," he says. Gradually, he focused on what was to become his field, the ecology of organisms that inhabit the meeting ground between sea and land. After graduate school Paine won a fellowship at the Scripps Institution of Oceanography in California, one of the world's major centers of research on the sea. After that he obtained a post as an assistant professor of zoology at the University of Washington. Today he is a full professor there and has served as a visiting professor abroad.

When I asked his advice about how a young person can prepare for a career in marine biology, Paine responded with some pointers that should help with any of the natural sciences.

First, he says, learn about a group of animals or plants—birds, for instance—as soon in life as you can. "The experiences at an early age will be remembered," he adds. Next, he advises, join nature organizations like the National Audubon Society, shell clubs, or home aquarists' clubs. Older experts in these clubs have much valuable information to share, and they can help greatly in encouraging a young person's abilities.

Lastly, Paine has a word for parents. "Let your children bring home snails and salamanders. Don't object if they want to skin the carcasses of birds found dead to learn about anatomy. This is the training ground. It doesn't matter what area of natural science eventually attracts a person. Start young."

7

Education

"I liked vacant lots," says Catherine M. Pessino when she describes how, as a young person growing up in the Bronx, New York, she became exposed to wildlife and nature. Her childhood was during the 1930s, and although the Bronx was not as crowded as today, it was still city, far from rural. Then, as even now, however, a surprising variety of wild creatures could be found in quiet neighborhood backyards, parks, and for that matter vacant lots.

Small snakes, a variety of birds including big ones such as hawks and owls, muskrats, frogs, turtles, raccoons, and of course an abundance of insects all live within the city limits of New York. Pessino has spent a considerable amount of time over the years trying to make New Yorkers aware of the wild animals living on their doorsteps.

Pessino is assistant chairwoman of the Education Department at The American Museum of Natural History, across from Central Park in Manhattan. The museum, one of the world's greatest, is not just a place that displays exhibits. It is a center of scientific research and education aimed at spreading understanding of the natural world far beyond the museum's own walls.

As a museum educator, Pessino is concerned with the full range of the institution's interests, ranging from folk arts to geology. Much of her work and background, however, are in biology, and her interests focus on wildlife—wildlife not only in exotic places, although she has been as far from the museum as India, but of her native New York City. Pessino supervised the establishment of the museum's Alexander M. White Natural Science Center. It is a portion of the museum that features exhibits representing the natural history of the city in which she was born and has lived and worked all her life.

A City Biologist

Pessino remembers always having a curiosity about nature. She liked to explore open spaces near her Bronx home, she says. Her interest in nature was encouraged by a teacher in the fourth grade who took Pessino and her classmates on field trips to the natural areas that remained in the neighborhood.

Sitting in her office on the third floor of the museum, Pessino related how she especially enjoyed watching birds. Surprisingly to many people, New York City can be a bird-watchers' paradise. Because it is located by the sea, it attracts many aquatic species. This is especially true during the migratory seasons. A large variety of songbirds also migrates through the city, which is located along a major aerial flyway—a route traveled by the flocks as they journey between northern breeding grounds and southern wintering areas.

While a biology major at a New York City college, Pessino often visited the American Museum to study the exhibits, especially birds. On graduation she hoped to find a job in the museum's Ornithology (bird) Department. Unable to find a job related to nature, however, Pessino applied to several hospitals for a position as a laboratory technician. But before she pursued this course further, however, one of her college biology teachers suggested that if she really wanted to work at the museum, perhaps she could find an opening in the Education Department.

Pessino enjoyed working with people, so she applied for and was given an assistant's job in education. Before long she had advanced and was teaching natural sciences to classes of schoolchildren. At the time, Pessino says, she was the only person in the department who had no background in education courses. She was a biologist who functioned as a teacher, not someone trained as a teacher who taught biology.

Today, Pessino explains, the situation is different at the museum and, from my own observations, at most other museums, zoos, aquariums, nature centers, and similar institutions that hire educators specializing in the natural sciences.

"When people come to learn at the museum," explains Pessino, "they look for expertise." Thus, she says, the museum prefers an education staff trained in scientific specialties rather than in formal educational techniques. Whatever their background, however, the members of the staff must be personally suited to instructing others.

Pessino progressed in the department until she became assistant chairwoman, second in command. Her superior in the department is the chairman and curator. As described in Chapter Four, a curator serves as the head of a department in a zoo or aquarium, and similarly at a museum. Pessino explains that under the policy at the museum, she could not become curator of the Education Department because a person in that position must have a doctorate. When she started at the museum, in the middle 1940s, a bachelor's degree was all that was expected, so Pessino did not undertake graduate study.

Educating Multitudes

Even so, Pessino is largely responsible for the daily operations of the department, which has a staff of twenty professionals plus about sixty volunteers. The department offers dozens of programs in and out of the museum; in a typical year these programs reach close to 400,000 people.

The museum's educational programs are tremendously varied. Here is a sampling of just a few:

—Gallery talks. Visitors to the museum can follow Education Department guides on tours of exhibit galleries. A talk given in a bird exhibit gallery, for instance, may be entitled "All About Birds"; another where sea animals are on display may be called "Whales, Penguins, and Seals."

—Lecture Series. Arranged by the Education Department, lectures are given for the public by museum experts on subjects such as "Animal Life in the Northeastern United States" and "Insects: Earth's Most Successful Animals."

—Local Field Trips. Each year lecturers from the Education Department escort a tour aboard an oceanic research vessel for people who want to see whales in New York coastal waters. This is one of several kinds of field trips, which also include nature walks in various parts of the city, offered by the department.

—College Courses for Teachers. New York City school teachers who wish to increase their knowledge of natural history can take courses offered by the museum's Education Department for college credit. An example: "Conservation of Wildlife," given by a lecturer from the Education Department who specializes in zoological subjects.

—Films. The department has programs featuring high-quality films on natural history. One that was being shown at the time this book was being written was "Secrets of an Alien World," A close-up look at insects, this film won an award at the 1982 Audubon International Film Festival.

—Workshops for Young People. "Understanding Animal Behavior" and "Birding for Beginners" are typical of the workshops given by members of the department staff for young people who wish to enroll in them.

—School Programs. Formal programs for school classes are held both in the museum and at schools. Department staffers use exhibits in the museum, and classroom demonstrations with specimens, to illustrate programs on topics such as "Shark Anatomy" and "Introduction to the Animal World."

Many of the programs are held at the Alexander M. White Natural Science Center, an exhibit area the size of one floor of a small house. Pessino developed the center and served as its supervisor when it opened in 1974. The purpose of the center is to teach people about nature using examples not from the wilderness but from the City of New York.

"Everything in the Center can be found within New York City," explains Pessino. "In some exhibits live animals are the focal point. For example, ants live in the crack-in-the-sidewalk exhibit; invertebrates and fish from a lake in Central Park are in the freshwater exhibit." Another exhibit, which perhaps takes Pessino back to her childhood days, is entitled "A Vacant Lot Is Not Empty." It shows the insects and other small animals that can live in vacant lots, together with the plants that make up their surroundings.

The education programs at the American Museum are so respected that Pessino was sent to India on a project supported by the Indo-U.S. Subcommission on Education and Culture. There she discussed new developments and techniques in her field with the education staff at the National Museum of Natural History in New Delhi, one of the leading museums in Asia.

Zoos, Aquariums, and Nature Centers

Jobs in education that put a person in contact with nature and wildlife are not so abundant as, say, those for zoo keepers. Even so, however, most major zoos and aquariums, plus many museums, nature centers, and sanctuaries, have education programs and need people to operate them.

By and large the programs offered are similar to those at the American Museum. Educators at the Mystic Marinelife Aquarium in Connecticut, for example, visit schools in their area to lecture on marine life and conservation of the seas. The aquarium has programs in which students can take part in field trips aboard research vessels. Special classes are held for students at the aquarium as well.

An important part of an educator's job at a zoo, aquarium, museum, or similar institution is preparing written materials for school teachers describing how they can best use the resources offered by their institutions. Often these materials are linked to major exhibits.

The Bronx Zoo, for example, has a sprawling exhibit complex known as Wild Asia. Visitors board a monorail train and travel through a series of outdoor exhibits where creatures such as Asian elephants, Indian rhinoceroses, and Siberian tigers roam seemingly free, but really confined by hidden barriers.

To assist schoolteachers who bring their students to Wild Asia, the zoo's Education Department produced a special teachers' kit. It contained materials for use both in the classroom, during discussions of Asian wildlife and wildlife habitats, and while visiting the zoo.

Teachers were given a set of instructions on how to use the kit and make the most out of the trip, a series of pamphlets on various aspects of Asian wildlife, a poster featuring vanishing species of Asia, audio-visual materials, activity sheets for students, and a paperback book called *Guide to Wild Asia*. The book was written by Douglas L. Falk, a member of the Education Department. In it he covered the environment, people, history, culture, and wildlife of the continent. In effect, although he was an educator, Falk was serving as a nature writer as much as any professional journalist.

Educators in the types of jobs described here often have a chance to act as nature writers. An educator at a nature center, for example, may write a pamphlet for visitors about birds found in the neighborhood. An aquarium educator may write fact sheets about aquatic creatures such as whales, seals, and penguins. Education departments sometimes are responsible for issuing regular newsletters or writing the texts for labels on exhibits. Films, slide shows, and posters may also be produced by the education staff. Educators at the Mystic aquarium even take their turns as actors in dramas designed to teach elementary school pupils about sea animals. The staffers dress in costumes representing creatures like sea anemones and horseshoe crabs, and carry on skits depicting the life-styles of these animals.

Sometimes an education department is so geared to communications that it is combined with a public relations department. Like the one I supervised for the New York Zoological Society, it then issues press releases and other information materials for the press, radio, and television.

The Future for Educators

Many of the people who serve on education department staffs are just beginning in their careers. The payment for regular staff members is often low—sometimes less than that of people in the scientific departments of institutions. At the same time, however, education staff members, especially those with graduate degrees in the sciences, have the chance to advance. Education curators are as highly paid and regarded as those of other departments. Sometimes a post as a curator of education for a zoo, museum, or aquarium can be a stepping stone to the job of director. Many nature centers, moreover, have directors who came from their education staffs.

8

Administration

When Robert F. Scott graduated from high school in 1939, he thought raising pheasants on a game farm would be an enjoyable way to earn a living. Game farms produce pheasants, quail, and similar birds for hunting preserves and state wildlife agencies, which release them to be hunted. Scott liked to hunt and work with animals, so he thought game farming would suit him.

Perhaps it would have. Game farming can be interesting and sometimes profitable. But Scott never became a game farmer. If he had, very likely his life would not have been as interesting. Scott has been involved in wildlife conservation from the ground up. Few people have served in as many different wildlife jobs as Scott, who throughout his life has moved along to positions of increasing responsibility.

118

Along the way, he has seen much of the world. Scott has mushed sled dogs through the snow-covered Alaskan wilderness while protecting reindeer herds from wolves. He has headed major programs for the Fish and Wildlife Service and National Marine Fisheries Service. He has lived and worked in Switzerland for an international conservation organization, the well-being of rare wildlife around the globe his concern.

Scott's specialty has become the administration of programs and organizations involved with wildlife. Like all other organizations and businesses, those working with wildlife need someone to run things. Budgets must be established, new projects planned, meetings scheduled, and the activities of many people coordinated. Scott has combined a background in biology and the outdoors with the ability to handle such tasks.

Administrators like Scott usually have long experience working with wildlife directly before becoming executives, which, in effect, is what they are. All of the careers described in this book can be stepping stones to administrative posts.

"Wildlife is a competitive field," says Scott. "Don't be afraid to start at the bottom."

A zoo keeper, for instance, does no administration. A curator does some. A zoo director's time is nearly all spent on it. With the right education, lots of ability, plus luck, a keeper can rise through the ranks to curator and perhaps director. The higher on the executive ladder however, the less direct contact the keeper has with animals.

Some of the people mentioned in earlier chapters do in fact perform administrative tasks. E. Hugh Galbreath of Remington Farms administers that institution. Catherine M. Pessino of the American Museum helps administer the programs of the museum's Education Department. Both, however, are still also involved in "hands-on" work in their fields—Galbreath as a wildlife manager, Pessino as an educator. Scott, on the other hand, has for many years been purely an administrator, often inspecting work in the field but no longer doing it.

Scott says he never actually set out to rise to the top levels of wildlife administration. "I did not have a specific goal," he says, "but I did have a sense of direction."

Throughout his career, Scott developed new talents and abilities, and sometimes gambled by taking on new challenges when he could have remained in positions that were comfortable. "The more I had to offer," he says, "the more opportunities opened up for me. Learning to read and write well helped."

Flying Squirrels and Crows

Like many other people who found careers with wildlife, Scott's interest in wild animals and nature in general goes back to childhood. He was reared in the suburbs of Boston, Massachusetts, then more rural than they are today. "I had close contact with the wildlife of the area," Scott says, although at times he yearned for places to

explore that were more remote than the fields and woods of the suburbs.

Among Scott's memories are tipping over a stone wall and finding a den with a family of weasels, and raising young crows that had fallen from the nest.

"I had read about flying squirrels," he tells, "but had never seen one." Finally he managed to shake one out of a tree and catch it. "I brought it home, and that night it had a litter of young," Scott says. "So I kept the squirrel and raised the young." When the squirrels were large enough to live on their own, Scott let them go. "I had a liberation ceremony for them," he says.

A big influence on Scott was one of his teachers in junior high school. She was deeply interested in nature and encouraged her students to make plant collections and go bird-watching. He also stresses that he read all he could about wildlife and the outdoors.

As a young outdoorsman Scott thrived on hunting and fishing. He trained his own bird dogs and hunted with them in the field. "I also started a rod and gun club when I was in high school," he says. His outdoor skills, which later in life served him well, were sharpened when he had the chance to attend a boys' camp in the deep woods of Maine.

Even so, Scott really did not expect he would have a career working with animals. "Although I had read a book about a boy who worked on a wildlife preserve," he says, "and wished I could have that type of job, I didn't know how to prepare for it or what was available."

So when he graduated from high school—he was only sixteen years old—the best he could think of was a job "raising pheasants." Scott found that the University of Massachusetts had a two-year program to train wildlife technicians, which qualified them for game farm jobs. He went to the school and investigated the course. A wildlife professor there interviewed him and, after listening to the young man, convinced him that there were other possibilities for wildlife careers. The professor suggested he enroll in a four-year college wildlife program and earn a degree.

Scott enrolled in the University of Maine and majored in forestry and wildlife conservation. The courses he studied gave him a sound background in zoology and ecology. He also found the chance to participate in field research.

The subject of Scott's research project was the ruffed grouse, a game bird very abundant in Maine. Scott focused on the mating behavior of the grouse, which is about the size and shape of a small chicken, with a feathery black collar—the ruff—on its neck. During the mating season, in spring, the male grouse puts on a spectacular display. Perching on a fallen log, the grouse rapidly beats with air with its wings. They make a loud, booming noise, known as drumming. The sounds made by the grouse thunder through the woods like the beats of a great bass drum, drawing females for mating with the male. Scott observed the types of surroundings in which the males chose drumming logs and compared these habitats with those preferred by the grouse in other seasons. He also mapped the locations of all drumming logs he found in a particular section of the forest.

Scott enjoyed this project and, he remembers, "I got good marks for it."

The summer before his senior year in college, Scott found the chance to work for the federal government in the type of big woods he had dreamed about as a boy. The United States Forest Service advertised for fire lookouts to work in the national forests. Scott was hired as a lookout in the vast Whitman National Forest of Oregon. He stood watch over the forest and, when necessary, worked as a "smoke chaser." He had to go by himself into the forest with a pack of food and tools to put out small fires after he spotted their smoke. Together with his college studies, the forestry job gave him experience at wildlife biology and managing wildlife habitat. It would serve him well when his career began in earnest.

Before Scott could begin to work with wildlife, however, he had to go to war. During his college years World War II raged. After Scott graduated from college, he entered the United States Army and was commissioned as a second lieutenant in an antitank unit.

Even as he prepared for war, however, Scott was thinking about his return to civilian life. He realized that to carry on the type of wildlife work he wanted, he would need a graduate degree. So he arranged with authorities at the University of Maine for graduate study after the war. He never returned to Maine, however, because he became caught up in the wild magic of Alaska and the northwest.

Scott's attachment to Alaska began in the Army, when he was sent to the Aleutian Islands. The Aleutians, bleak, barren islands that extend west from the Alaskan mainland, are to some people the windswept ends of the earth—lonely, brutal, and depressing. Scott found them a "wonderland," with their vast herds of seals, colonies of sea otters, and mobs of seabirds.

When the war ended, Scott began looking for graduate schools in the northwestern part of the country. He picked Oregon State, which provided him with a fellowship to study fish and game management, with a minor in zoology. Scott earned his master's degree in two years.

Next he pondered how to get back to the Aleutians and Alaska. His military background, which included Arctic and mountain survival training, should add to his credentials, Scott reasoned. Hearing that the Fish and Wildlife Service director for the Alaskan region would be visiting Oregon, Scott arranged to meet and get to know him. But, he found, there were no job openings in the region.

Meanwhile, to be safe, Scott sent dozens of letters to state fish and game agencies, federal government agencies, and other organizations that might have job openings suited to him. And, indeed, he received offers of positions with some of them. At the last minute, however, word came that the Fish and Wildlife Service needed a wildlife biologist to carry on a project for the summer. The task was to travel the Innoko River, banding ducks and studying the habitats there to see if the area would make a good national wildlife refuge. Scott had to decide between offers of permanent jobs and a temporary job in the area he wanted most. Gambling, he chose Alaska and a job for only a summer.

Traveling to Anchorage, Alaska, Scott contacted a Fish and Wildlife biologist who was to start him out on the project. The next day he and the biologist flew to a religious mission for Alaskan Indians in the wilderness, where the biologist had stored a boat a few years before. Their reception was not friendly. Alaska state game wardens had arrested some of the Indians at the mission for trapping too many beaver. The clergy running the mission were angry over the arrests, which they thought unfair, so they presented Scott and the biologist with a bill asking a large fee for storage of the boat. It asked more money than they had, Scott recalls, so, he explains, "We talked our way out of it."

With an Indian guide Scott motored 400 miles upriver into the wilderness, observing the environment and the wildlife living there and banding waterfowl. It was, he remembers, a good summer, his hopes, for a time at least, coming true. At summer's end, however, the project was completed.

One day, while Scott was on his way to a village store to buy supplies, a trader stopped him. The trader had been listening to his radio communications set and heard a telegram that was being relayed to Scott. It was offering a job teaching parasitology—the study of animal parasites—at a university in one of the lower 48 states. A course in parasitology qualified him for the post.

Scott decided that he had to leave to accept the job, which also involved studies of freshwater game fish. As he passed through Anchorage on his way out of Alaska, however, fate took a hand. Fish and Wildlife had a low-paying but permanent job for him. It was as a wolf hunter.

Over the years there has been considerable controversy over the government's policy of protecting deer, reindeer, and moose by thinning down the number of Alaskan wolves when they get too numerous. There are pros and cons on both sides. This is not the place to debate the question. The important point is that the job enabled Scott to stay in Alaska in a wildlife post.

By dogsled and airplane—Scott learned to fly—he patrolled herds of moose and caribou. It was a tough outdoor job that expanded Scott's wilderness experience. After a year a new opportunity arose. Scott became a Fish and Wildlife biologist, carrying on studies of animals such as mountain sheep and caribou so they could be managed more effectively. He published reports of these studies in scientific journals, which added to his qualifications.

Later, Scott headed the Alaska Cooperative Wildlife Research Unit at the University of Alaska, where he also became an associate professor. Several universities have such units, backed by Fish and Wildlife. The unit provides a chance for the university's graduate students in wildlife ecology and management to undertake field research. The post gave Scott a taste of administering scientific research programs, something on which he would build later in life.

Scott himself wanted to study for a doctorate. He saved up vacation time and started a program at the University of British Columbia, in Canada, where he studied off and on for several years. Due to the pressures of his job, he never found time to complete the degree, but he considers the experience one of the most important in his life.

In 1961 Scott finally left Alaska for another Fish and Wildlife post. It was at the Patuxent National Wildlife Research Center in Maryland. The center is the site of major research programs, ranging from the effects of pesticides on birds to the ecology of endangered species. It is on a sprawling area of abandoned farms not far from Washington, D.C. Scott's job was to study the effects of introducing a plant called multiflora rose, good for wildlife, on the grounds. He also led visitor tours of the center. Other new types of experience were added to those he already had accumulated.

Scott spent four years at Patuxent, which exposed him to the administration of a large government installation. In 1965 Fish and Wildlife brought him to the agency's Washington, D.C., headquarters, from which, for two years, he performed a variety of important functions. One was to travel the country surveying federal lands, such as Indian reservations, to see how they could be better managed for wildlife. Another was to confer with state conservation agencies to see how effectively they could use assistance from the Fish and Wildlife Service. As he performed these tasks, Scott learned the details of how government conservation agencies operated—the basics of administration.

A Major Promotion

The chance to become a full-time administrator came in 1967. Fish and Wildlife named Scott as chief of its Division of Wildlife Refuges, one of the nation's most important wildlife conservation positions. The National Wildlife Refuge System began in 1903, when President Theodore Roosevelt set aside Pelican Island, off the east coast of Florida, as a haven for herons and egrets, then being slaughtered without control for their plumes. Today there are almost 400 national wildlife refuges, totaling more than 33 million acres of land and water. They extend not only over the 50 states but to regions overseas controlled by the United States—Baker Island in the southwestern Pacific, for example.

Land in the refuge system was either already owned by the federal government and declared refuge or donated to the government for wildlife conservation. The Union Camp Corporation, a timbering and paper company, for example, gave 49,000 acres of the Dismal Swamp of Virginia and North Carolina to the refuge system. The Nature Conservancy, a conservation organization, has donated land in several parts of the country to Fish and Wildlife as national refuges.

Refuges were originally set up mainly for waterfowl and other aquatic birds, largely because wetlands destruction was threatening them. Eventually, however, the refuges became places for the conservation of a wide variety of animals. The Key Deer National Wildlife Refuge in Florida, for instance, was established in 1957 for the tiny white-tailed deer of the Florida Keys. The National Elk Refuge in Wyoming was created to provide habitat for 7,000 elk in a herd wintering in that state's Teton Valley.

National wildlife refuges are not merely preserves. They are centers of wildlife management, where the land is manipulated to encourage various types of wildlife. A wide field of human activities, moreover, is allowed on many refuges. These range from timber cutting and hunting to environmental education and camping.

Until Scott took over the system, each refuge was operated independently, poorly coordinated with activities at others. The system had little in the way of long-term, organized planning. Whether or not a refuge was run effectively usually depended entirely on the abilities of its chief manager, not the overall system.

Scott brought a businesslike approach to the refuge system. A more purposeful direction, he realized, was needed, especially since the money and staffing available to the system were limited. If one looked at the refuge system as a unit, set goals for the entire system, and combined resources such as obtaining information, Scott believed, the operation could become a much more effective force in wildlife conservation.

Within two years Scott and his staff had developed a new method for operating the refuge system. Employees at the various refuges were trained to adapt to the new ways of doing things. Scott helped Fish and Wildlife pinpoint goals for the system as a whole, as well as for individual refuges. By surveying the different objectives and needs of the system, Fish and Wildlife officials could decide which to stress and supply with the most funds. By 1971 the refuge system had begun to operate according to the methods outlined by Scott.

New Directions

One of the principles that has guided Scott through his career is his belief that a person engaged in science should change focus every so often to prevent becoming stale. So after helping the refuge system reorganize, Scott spent a year setting up a new planning division for the Fish and Wildlife Services. He then began to think of leaving the service. He had purchased some land on the scenic San Juan Islands off the coast of Washington, and began to wonder about moving there, especially since a marine laboratory on the islands was looking for a new director.

Just about that time, as he was thinking of marine

conservation, a friend who worked for the National Marine Fisheries Service approached him. The friend headed a new division concerned with resource research—investigating the outlook for fisheries in the future. Why not join up and help plan the new operation? the friend asked. Scott accepted and joined the service. He remained until 1979, along the way becoming the acting chief of the service's Office of Science and Environment. The office, with a staff of 1,000 people, oversaw all the service's fisheries and marine mammal research. Scott did not conduct research himself but was concerned with supervising the agency's operations, much as he had with the refuge system.

A Job in International Conservation

Even with a full career, Scott was still ready for yet more challenges. His next aim was to enter the world of international wildlife conservation. When he heard that the World Wildlife Fund was looking for a new scientific advisor, he applied. Headquartered in Switzerland, the fund supports and promotes conservation in scores of countries by providing money for projects and encouraging public interest in wildlife and the environment. Scott did not get the job. It went instead to a former colleague of Scott's in the federal government, like him an able administrator with a long background in wildlife biology.

World Wildlife is closely associated with another conservation organization, based with it in Switzerland. It is the International Union for the Conservation of Nature and Natural Resources (IUCN). The IUCN is a union of conservation and scientific organizations from more than a hundred countries. Members are government agencies, scientific institutions, and conservation organizations. The IUCN is organized into six commissions, each with a special area of concern. One, for instance, is involved with national parks, another with rare species.

The IUCN's activities include monitoring what is happening worldwide in conservation and calling attention to problems and needs. It helps plan and promote conservation programs, and provides scientists to help carry them out with money provided by organizations such as World Wildlife, the New York Zoological Society, and the United Nations.

A Canadian wildlife administrator who once worked with Scott was head of the IUCN. Scott contacted him and mentioned he was interested in an international job. A year later the contact paid off. One of the IUCN's commissions—the Species Survival Commission— needed an executive secretary. Scott applied and was given the job on a trial basis.

It was while engaged myself in a project for the Species Survival Commission—it is the one concerned with rare species—that I met Scott. He was still in his trial period, but had already settled in and shortly after was given the post permanently.

Tall and white haired, Scott is, as a former head of Fish and Wildlife described him, "a thoughtful, intellectual man." Quiet, soft-spoken, but with ready humor, Scott looks much more like a professor than a wolf hunter and outdoorsman. It is only after spending some time with him that one grasps that he has had a lifetime of familiarity with the wilderness and creatures living there. If he leaves one lasting impression, it is that he is a gentle person, but that his kindness is backed by strength.

These qualities, in fact, have enabled him to be an excellent administrator. Scott can handle people, even when they are at odds with one another. As executive secretary of the Species Survival Commission, Scott must coordinate the efforts of hundreds of different scientists and conservationists, many of whom may have different ideas about what is best for wildlife.

I have watched Scott settle disagreements between people attending a meeting by quietly discussing the differences of opinion, clarifying the issues, and finding common ground. All the while, moreover, he keeps above the arguments himself, which allows him to help settle them. Whether a person is in the business of wildlife conservation or making automobiles, this type of talent is what makes a good administrator, whatever other experience he or she has.

Scott is also a good person with details. An important part of his job is to arrange for meetings of the commission on a regular basis. Because people who belong to the commission come from many countries, meetings are held in different parts of the world. Florida, Kenya, and India, for example. Scott must make sure all is in order well ahead of time—from places for members to stay and meet to coffee or tea during breaks in meetings. Well in advance, each member must be notified of the arrangements as well as the topics to be discussed. Members who want certain subjects brought up at the meeting contact Scott, who sees to it that these are on the meeting agenda. At the meeting itself Scott keeps minutes, which are later prepared as a report for each member. If any problems crop up, he must be ready to deal with them.

After meetings, the members return to their home countries and regular jobs. Scott goes back to his office in Switzerland, where he keeps the daily affairs of the commission in order between meetings. The commission advises on an endless stream of projects which deal with wildlife in any number of countries. Scientists and others working on these projects use Scott as their point of contact with the commission. Is one researcher in southern Asia running out of funds for a project? Is another having difficulties with the government officials of the country where a project is underway? Is another failing to perform the job required? All these details and a flood of others find their way to Scott's desk, where he must sort them out.

Without his own background in wildlife, Scott might have a difficult time even though his nature suits him to administration. But Scott understands the wildlife field, from the problems faced by a researcher working in a remote camp to those of someone running an agency. When all his talents are put together, he makes a top wildlife administrator.

Scott seems fond of remembering past actions that

suddenly years later helped him in his career. Even his uncompleted attempt to earn a doctorate assisted him, in a way he never expected when he was still attending classes at the University of British Columbia.

"I had to learn French and German to read wildlife writings in those languages," Scott recalls. "I never really had to use those languages in my practice of biology. But now with our commission's international membership, and with the commission headquartered in Switzerland—where French and German are official languages—they are important."

Appendix

ORGANIZATIONS YOU CAN JOIN
TO LEARN ABOUT NATURE AND WILDLIFE
(those listed have major publications for members)

American Forestry Association
1319 18 Street NW
Washington, D.C. 20036

American Littoral Society [coastal conservation]
Sandy Hook
Highlands, New Jersey 07732

The American Museum of Natural History
Central Park West at 79 Street
New York, New York 10024

Boy Scouts of America
North Brunswick, New Jersey 08902

Ducks Unlimited
P.O. Box 66300
Chicago, Illinois 60666

Friends of the Earth
124 Spear Street
San Francisco, California 94105

Game Conservation International
900 NE Loop 410
Suite D–211
San Antonio, Texas 78209

Girl Scouts of the United States of America
830 Third Avenue
New York, New York 10022

Izaak Walton League of America
1800 North Kent Street
Suite 806
Arlington, Virginia 22209

National Audubon Society
950 Third Avenue
New York, New York 10022

National Geographic Society
17 and M Streets NW
Washington, D.C. 20036

National Wildlife Federation
1412 16 Street NW
Washington, D.C. 20036

The Nature Conservancy
1800 North Kent Street
Suite 800
Arlington, Virginia 22209

New York Zoological Society
Bronx Zoo
Bronx, New York 10460

Sierra Club
530 Bush Street
San Francisco, California 94108

Smithsonian Institution
1000 Jefferson Drive SW
Washington, D.C. 20560

The Wilderness Society
1901 Pennsylvania Avenue NW
Washington, D.C. 20006

World Wildlife Fund
1601 Connecticut Avenue NW
Washington, D.C. 20009

Index

E

eagle, bald, 48–49
education, careers in,
    109–17
  education for, 111–12
  employment in,
    111–12, 115–17
  job responsibilities,
    110, 112–17
egrets, 12, 127
elephants,19–20,21–22,
    23, 26, 27, 29–35, 115
Environmental
    Protection
    departments, 10, 41,
    42

F

farmers' cooperation in
    conservation, 7
FBI, 36, 41, 49, 56
federal government, 7,
    10, 38, 40, 41, 123.
    See also specific
    agencies.
ferrets, black-footed, 22,
    24
field biology, careers in,
    19–35, 79
  definition of, 20–21
  education for, 24–25
  employment in, 22–23,
    25

job responsibilities,
    23–24, 32–35
Fish and Game
    Department, 41
Fish and Wildlife
    Service, 10, 13,
    22–23, 41, 48–49,
    50–52, 56–57, 77,
    119, 124–30
Florida, University of, 25
Forest Service, 10, 123
foxes, 12
Franklin Park Zoo
    (Boston, Mass.), 48
Friends of the Earth, 93

G

Galbreath, E. Hugh,
    12–17, 120
game wardens. *See*
    wildlife law
    enforcement.
Garibaldi fish, 62
geese, Canada, 5–7, 12
Georgia, University of,
    13
Glynco (Ga.), agent
    training school at, 57
grouse, 122

H

habitat, loss of, 8, 21
Harvard University, 105,
    106